APPROACHES TO IMPLEMENTING
SOLID WASTE RECYCLING FACILITIES

APPROACHES TO IMPLEMENTING SOLID WASTE RECYCLING FACILITIES

by

Marc J. Rogoff and John F. Williams

np NOYES PUBLICATIONS
Park Ridge, New Jersey, U.S.A.

Library of Congress Catalog Card Number:94-3865
ISBN: 0-8155-1352-6
Printed in the United States

Published in the United States of America by
Noyes Publications
Mill Road, Park Ridge, New Jersey 07656

10 9 8 7 6 5 4 3 2 1

Library of Congress Cataloging-in-Publication Data

Rogoff, Marc Jay.
Approaches to implementing solid waste recycling facilities / Marc J. Rogoff and John F. Williams.
p. cm.
Includes bibliographical references and index.
ISBN 0-8155-1352-6
1. Recycling (Waste, etc.) I. Williams, John F., 1955- II. Title.
TD 794.5.R64 1994
363.72'85--dc20 94-3865
CIP

ABOUT THE AUTHORS

Marc J. Rogoff is a Vice President with HDR Engineering, Inc., located in their Tampa, Florida office. He is responsible for management of the firm's solid waste management practice in the southeastern United States. He has experience in both government and industry for the implementation of solid waste management programs. Dr. Rogoff has a Ph.D from Michigan State University, an MBA from the University of Tampa, and M.S. and B.S. degrees from Cornell University. He is an active member of several solid waste professional organizations, particularly the Solid Waste Association of North America, which he serves as an international Board Member, the National Solid Waste Management Association, and the Municipal Waste Management Association. He has published widely in the solid waste field in numerous solid waste periodicals.

John F. Williams is a Vice President and National Solid Waste Program Manager with HDR Engineering, Inc., located in their White Plains, New York office. He works with HDR's thirty offices in the development of alternative technologies and innovative public education programs. He was a pioneer in the area of recycling and environmental education for children ages two to four. His work with public television's MISTER ROGERS on the nation's first major program designed to teach preschoolers about recycling and conservation earned five national awards including the National Recycling Coalition's Fred Schmidt Award for Outstanding Corporate Leadership, USEPA Administrator Reilly's Award for Environmental Education and Keep America Beautiful, Inc.'s National Award of Excellence. Mr. Williams holds a BSLA degree from West Virginia University. He studied psychology and communications at Fairmont State College and received professional development training at Harvard University's Graduate School of Design. He is a technical advisor on alternative technologies and Clean Air Act Amendment issues to the U.S. Conference of Mayors, Municipal Waste Management Association. He is also an active member of the Solid Waste Association of North America and the National Recycling Coalition. Mr. Williams writes frequently for MSW Management, World Waste, American City and County as well as Solid Waste Technologies Magazine.

ACKNOWLEDGEMENTS

This book was made possible by the cooperation and participation of many friends throughout the solid waste industry. We extend our thanks to all who took their often limited times to chat with us and discuss our observations and conclusions. We would also like to extend our many thanks to our colleagues at HDR Engineering, Inc. In particular, we would like to acknowledge the assistance of Peter Lieben who edited much of the earlier drafts of this book. His many suggestions made the document undoubtedly more readable.

We would also like to extend our thanks to our families who had to contend with our sometimes long hours spent away from family pursuits. Without their understanding and encouragement this book would never be written.

PREFACE

Over the past few years many communities throughout the United States have implemented recycling programs. These have included residential curbside recyclables and yard waste collection programs either using municipal work forces or through agreements with private haulers. The most successful of these programs have included ongoing public education and information programs tailored to the community's demographic and cultural conditions. Economics has also been a critical concern in the development of these programs as communities face evermounting costs to provide municipal programs to address other pressing urban needs.

Both of us serve as solid waste consultants to many communities and have written widely on the issue of integrated solid waste management. However, in our travels it appeared that many public agencies, which had responsibilities for implementing solid waste recycling programs, needed a standard text illustrating practical approaches for recycling solutions. It is our hope that this book will provide needed guidance to these public officials.

Marc Rogoff
Tampa, Florida

John Williams
White Plains, New York

April 1994

ACRONYMS AND ABBREVIATIONS

A/E	Architect/Engineer
API	American Paper Institute
ASTSWO	Association of State and Territorial Solid Waste Management Officials
BAN	Bond Anticipation Note
CPRR	Center for Plastics Recycling Research
CFR	Code of Federal Regulations
DEP	Department of Environmental Protection
e.g.	For example (exempli. gratia)
EPA	U.S. Environmental Protection Agency
et al.	And others (et alia)
ft.	Foot (feet)
GNP	Gross National Product
G.O.	General Obligation
HDPE	High Density Polyethylene
HHW	Household Hazardous Waste
i.e.	That is (id est)
IDB	Industrial Development Bond
lb	Pound
LDPE	Low Density Polyethylene
MRF	Materials Recovery Facility
MSW	Municipal Solid Waste
NA	Not Available
NIMBY	Not-In-My-Backyard
NSWMA	National Solid Waste Management Association
No.	Number
OCC	Old Corrugated Cardboard
ONP	Old Newspaper
O & M	Operations and Maintenance
OWP	Office Waste Paper
p.	Page
pp.	Pages
PCD	Pounds Per Capita Per Day
PET	Polyethylene Terephthalate
PP	Polypropylene
PS	Polystryene
PVC	Polyvinyl Chloride
RCRA	Resource Conservation and Recovery Act
RFP	Request-For-Proposal

RFQ	Request-For-Qualifications
RMP	Residential Mixed Paper
SWDA	Solid Waste Disposal Act
SWANA	Solid Waste Association of North America
t	Ton(s)
tpd	Tons Per Day
tph	Tons Per Hour
tpy	Tons Per Year
U.S.	United States
WTE	Waste-To-Energy
yd	Yard
yr	Year

Signs and Symbols

>	Greater Than
<	Less Than
#	Number
′	Feet
"	Inches
$	Dollars

NOTICE

To the best of the Publisher's knowledge the information contained in this publication is accurate; however, the Publisher assumes no responsibility nor liability for errors or any consequences arising from the use of the information contained herein. Final determination of the suitability of any information, procedure, or product for use contemplated by any user, and the manner of that use, is the sole responsibility of the user.

The book is intended for informational purposes only. The reader is warned that caution must always be exercised when dealing with chemicals contained in, or procedures involved with, recycling procedures. Expert advice should be obtained at all times when implementation is being considered. Mention of trade names or commercial products does not constitute endorsement or recommendation for use by the Publisher.

The reader must contact the appropriate legal counsel and regulatory authorities for up-to-date regulatory requirements, their interpretation, and implementation. The book is sold with the understanding that the Publisher is not engaged in rendering legal, engineering, or other professional service. If advice or other assistance is required, the service of a competent professional should be sought.

TABLE OF CONTENTS

CHAPTER **PAGE**

INTRODUCTION

Many Americans believe that recycling will pay for itself. In fact, the claims that new industries and related job opportunities will be created could be viewed as an implied promise, yet to be fully realized. Certainly, more and more material recovery facilities are being opened and material manufacturing industries are retooling to accept recyclables. However, as communities strive to broaden their recycling efforts, the associated costs have risen and are often not offset by the development of a new industry.

Figure 1-1 shows the most rapidly growing costs facing cities and counties today. With recycling and solid waste management costs outpacing those associated with crime, education, child care, infrastructure and welfare, it is easy to see why public officials would be looking for greater results from recycling. Many communities are faced with difficult choices when it comes to expanding or even maintaining current recycling efforts. How will they pay for the service while state and federal financial support disappears? The answer often comes down to a choice between other critical services like police and fire protection or even street cleaning and recycling. It is a difficult choice because pressure is growing to force recycling programs to be efficient and to some extent, self sustaining.

Although scrap materials have been traded since the time of the earliest settlements in the United States, institutionalized recycling of materials from the solid waste stream can be considered in its infancy. Since 1988, more than 5,000 communities throughout North America have initiated community-wide recycling programs. Most, if not, all of these programs were begun because of a statewide legislative policy directive. Others were initiated by civic-minded citizens who were concerned with the waste of often scarce resources. Still other programs were begun by communities whose political leaders saw recycling as a panacea to forestall the need to site politically unpopular facilities for the collection, transport, processing or disposal of solid waste. However, with the maturing of the recycling industry in the United States, most communities are now recognizing that there is "no gold in recycling" and there are no quick fix solutions to the solid waste issue.

The authors' objective in writing this book is to provide a basis for the development of judgement that will be critical in recycling decision-making.

1.1 Practical Approaches to Recycling

There is growing concern that many recycling decisions were driven by influences other than economics and the strict interpretation of environmental impacts. High expectations have lead to disappointment in communities that expected recycling to take the place of the local landfill or waste-to-energy facility. The National Recycling Coalition reports that, after nearly a decade of public attention focused on recycling, the total elimination of recyclable materials from the waste stream destined for disposal is far from complete. In Figure 1-2, recovery rates for paper and containers are shown for 1992. While progress has been made, the results show that there may be limits to what can be achieved through current methods of recycling.

1.2 Responding to Public Needs and Concerns

Given uncertainties relative to cost, markets and net reduction in waste to be disposed of, what then is the proper role of recycling in current solid waste management? What new technologies must be developed to efficiently collect recyclable materials? What can be done to open up the markets for these materials and return them to the manufacturing sector as raw materials for new products?

These are some of the most important questions and challenges that will face current and future recycling coordinators and other solid waste professionals. Recycling and solid waste practitioners who witnessed the birth of recycling for political, philanthropic or romantic reasons will soon see recycling overcome by economic, environmental and technical realities. Recycling programs will survive if they are truly responsive to public needs and provide legitimate environmental justification.

The intent of this book is to address the basic educational needs of individuals who decide to devote a portion of their careers to the development or expansion of recycling programs. Interviews were conducted with prominent professionals in the solid waste and recycling industries as well as pioneers in public education. Included were Marsha Rhea, Executive Director of the National Recycling Coalition (NRC); Lanier Hickman, Executive Director of the Solid Waste Association of North America (SWANA); David Gatton, Executive Director of the Municipal Waste Management Association (MWMA); and Roger Powers, President of Keep America Beautiful, Inc. (KAB). Most were in agreement that public expectations do not match the realities that face recycling. There is a growing challenge in directing public dollars to recycling efforts.

The constantly changing nature of the market place and public policy on federal, state and local levels has done little to secure the long term future of recycling. Issues related to the control of recyclable materials from the household to final disposition; the re-authorization of the Resource Conservation and Recovery Act (RCRA) which could attempt to mandate recycling rates, minimum recyclable materials content, advance disposal fees (such as deposits), eco-taxes (artificial fees designed to discourage the purchase of specific materials or products), and/or manufacturer take back programs that, in essence, force manufacturers to "recycle" or dispose of waste materials themselves; emerging regulations that address recycling worker health; and changes in mandatory recycling goals will shift the ground rules for recycling without truly addressing the basic issues that will make or break recycling. Recycling and solid waste professionals will be expected to sort out new legislated developments while maintaining day-to-day operations of collection, processing and marketing programs. In the end, their programs will be judged upon rate of return for investment required.

Recycling and solid waste manager jobs are growing in complexity and could be increasingly challenging. Figure 1-3 uses the image of a hurdler to make this point. Unfortunately, even world class track athletes train for a consistent hurdle height. Recycling and solid waste professionals are constantly facing changes beyond their control that effectively increase the height of the hurdles in their paths. Constant training and access to the latest technical and market developments are the only way professionals can continue in the recycling race.

1.3 "It's the Economics"

When Bill Clinton was running for president in 1992, the press made a big deal of a poster that hung in the campaign "war room". The words on that poster became famous and helped to focus candidate Clinton on a strategy that won him the presidency. The poster reminded key staff that, "It's the Economy...Stupid". Current and emerging recycling professionals will find that it's the economics of their programs that cause them to succeed or fail. Their abilities to guide public or private programs through decision-making processes that maximize recycling results while minimizing cost will be key.

1.4 Flexibility

In order to make the right decisions, recycling and solid waste professionals will need training that can help them:

- Gain public or employee support and participation;
- Yield materials that meet market specifications without unreasonable investment;
- Minimize collection costs; and
- Make use of processing approaches that are efficient and reliable.

Figure 1-1

MAINTAINING PUBLIC SERVICES: MOST RAPIDLY GROWING COSTS

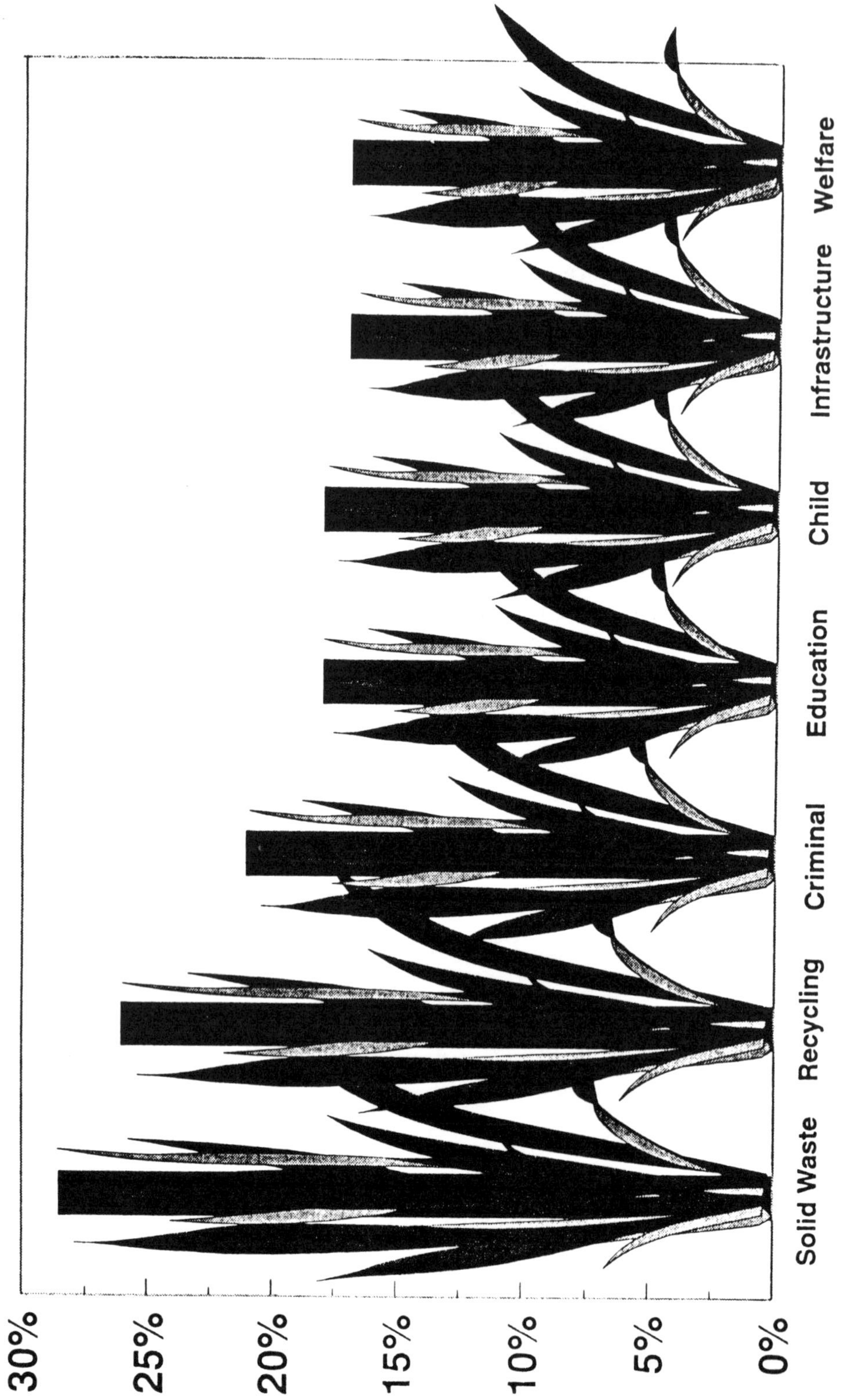

Source: American City and County, 1993

Figure 1-2

1992 RECOVERY RATES FOR PAPER AND CONTAINERS

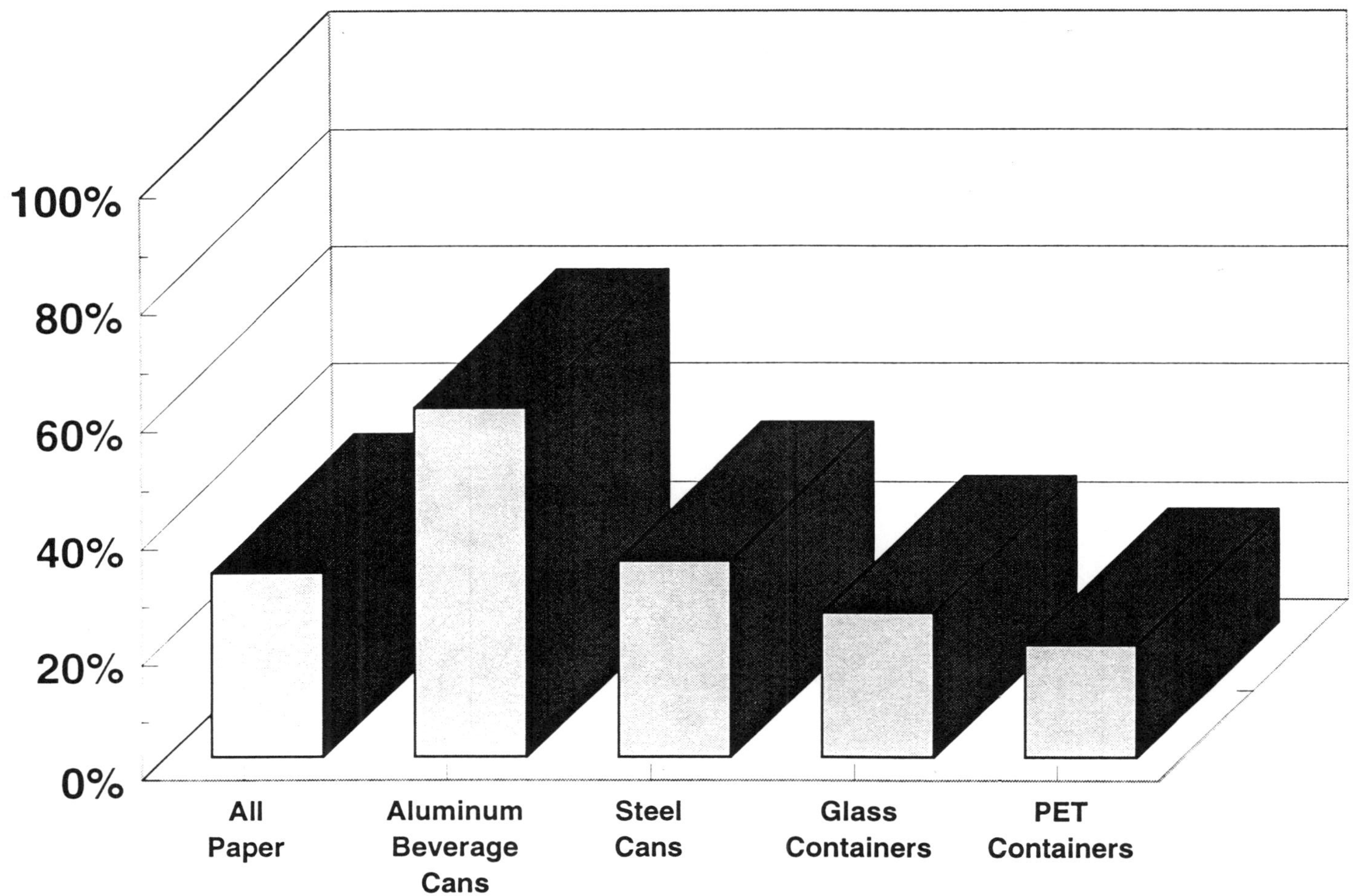

Source: National Recycling Coalition, 1993

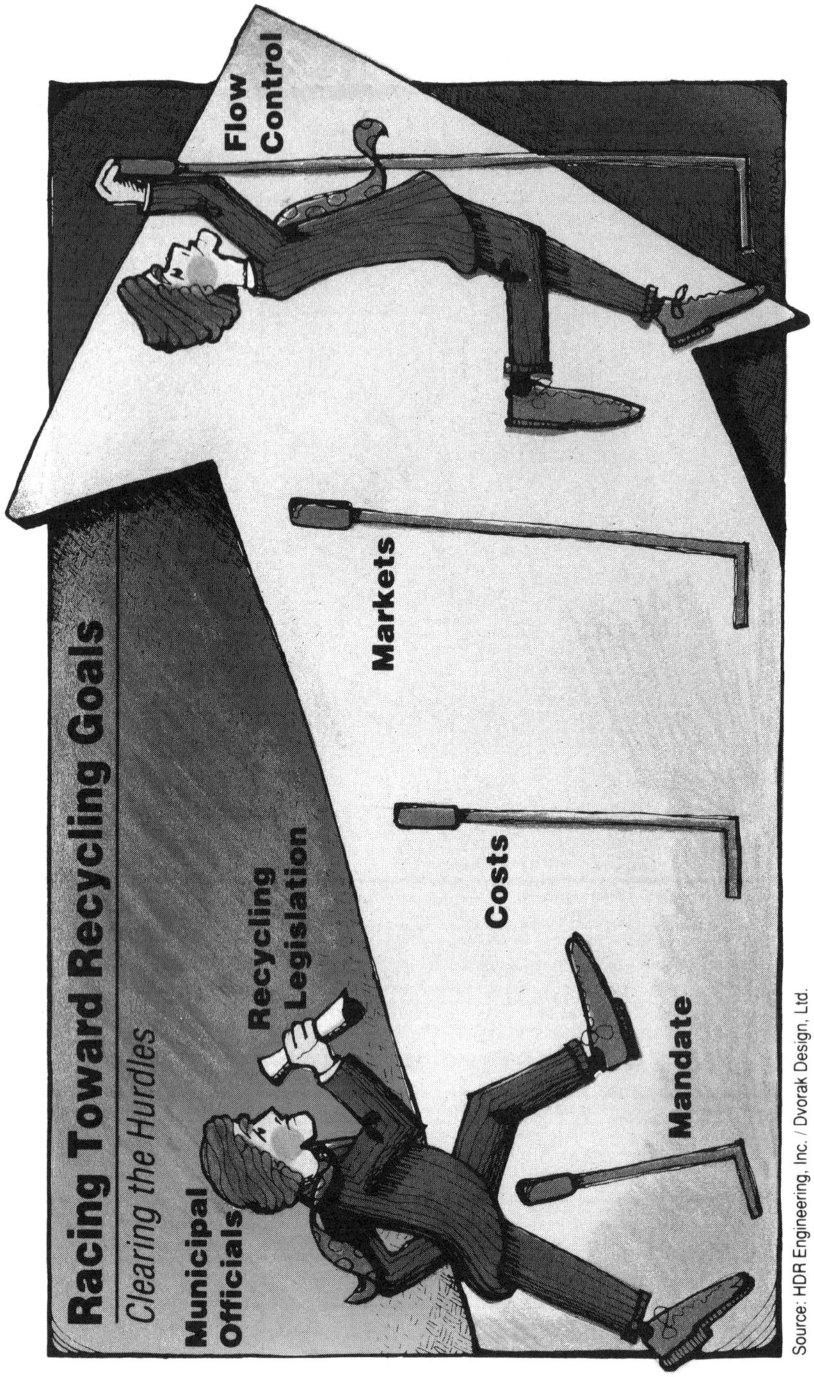

Source: HDR Engineering, Inc. / Dvorak Design, Ltd.

Most importantly however, will be the need to plan for flexibility in every aspect of their programs. While, in the end, it's the economics of the programs that will make them work, recycling and solid waste professionals must be aware of and plan for the potential impact of regulation and government intervention in their programs and markets.

1.5 Related Training

Individuals embarking on formal undergraduate training in preparation for pursuing a career in solid waste management that includes recycling should consider the basics provided in this volume. Also, however, they should look into other preparation needed to fulfill their career goals. Lanier Hickman of SWANA advised that students on an undergraduate level should develop a background in engineering. Depending on the specific area of interest in recycling, mechanical engineering or environmental engineering could provide strong preparation. If not pursuing an engineering degree, study should include emphasis in the sciences, particularly chemistry, microbiology and biology. Information technologies would also be helpful with strong orientation toward computer sciences. Given the importance of the public sector's role in most recycling programs, government or public administration and political sciences will be very helpful.

Graduate level training in solid waste management is becoming more and more available at institutions like the University of Wisconsin, Cornell University and the Rochester Institute of Technology. These and many other similar academic programs provide a wider focus on solid waste technology, systems and practices. Professional associations like the Solid Waste Association of North America have developed "certification" programs for individuals seeking to switch between related industries.

1.6 Career Paths

Given the emphasis of this book on training individuals that aspire to become involved in recycling or solid waste management, it is appropriate to dedicate a few words to the potential employment opportunities that may exist for properly trained and experienced individuals. There are seven primary paths that could be pursued from a career point of view. They include the following:

- Regulatory;
- Program Implementation;
- Consulting;
- Equipment supply;

- Service provision; and
- Professional association support.

1.6.1 Regulatory

The regulatory path would involve work most likely at the state level generally via environmental agencies. The development of state policy and environmental regulations entails a knowledge of public administration and environmental sciences. Regulators with a working knowledge of economics and interface with technology and materials marketing would, in some cases, yield practical requirements that may be more feasible on a local level.

1.6.2 Program Implementation

Program implementation activity generally occurs at the local level in either a public position charged with municipal recycling activities or a commercial position responsible for making recycling work in a for profit setting. In either case, project management skills will be important. Time and budget will be prominent issues as will be the ability to provide consistency in recyclable product quantities and quality. Persuasive skills will also be important given the interface required with policy makers and individuals involved in providing program funding.

1.6.3 Consulting

Consulting services generally offered through specialty engineering or planning firms requires knowledge founded in technology, economics and/or regulatory affairs. Consultants are most often used to provide input in program planning and evaluation; collection or processing technology selection and testing; economic analysis; financing opinions; facility design and permitting; construction monitoring; and/or system procurement. A key difference between consulting in this arena and other traditional engineering services is the role that the professional plays in bringing public sector needs together with private sector services. In general, recycling and solid waste management are areas where true public-private partnerships are possible. The process by which agreements are reached is unique when compared to traditional engineering or public sector roles and requires a strong technical background as well as a working knowledge of business and legal issues.

1.6.4 Equipment Supply

Equipment suppliers are often interested in staff that can design material handling equipment or help in selling their products. Generally, engineering backgrounds are most helpful in pursuing this career option.

1.6.5 Service Provision

Service providers like WMX Technologies (formerly known as Waste Management of North America), Browning Ferris Industries, Laidlaw and many other smaller companies, typically provide recyclables and/or solid waste collection, processing, materials marketing and residue disposal. They employ significant staff to support their operations and sales functions. A few companies are invoked in the global market and have resources used abroad to develop recycling programs. There are also joint projects between service providers and individual companies that could together, create or sustain major materials markets. Service providers are generally interested in a working knowledge of solid waste management and recycling. They are very interested in materials marketing experience, skills in project accounting and operations management.

1.6.6 Professional Association Support

Professional associations also hire individuals with backgrounds in recycling and solid waste management. Groups that represent elements of the industry be they governmental advocates, equipment or service provider associations, groups that represent recycling coordinators or solid waste managers, and environmental lobbying organizations are all interested in individuals that have specific knowledge tied to waste management and recycling.

1.7 From Art to Science

Unlike most other technically based fields, recycling and solid waste management can be as much an art as science. There is no doubt that factual assessments of recyclables and waste availability, project economics, equipment performance data, detailed facility design and environmental analyses provide a major portion of the basis for decision making. However, public perception, political winds, and regulatory ramifications often become the driving force behind recycling and solid waste management programs. The recycling and solid waste manager must be equipped to anticipate and understand the emotional influences, while prepared with factual information to make decisions that, in the end, best serve the community or business providing funding for the program. The balance of this book addresses both art and science aspects of recycling. From the factors that influence the community's true recycling potential, including institutional issues that limit community latitude, to strategy for public involvement, the equipment selection process and alternatives for collection, when cloaked in flexibility, the combination of lessons should yield programs that offer the best possible economics and associated odds for success.

-2-

CALCULATING MAXIMUM RECYCLING POTENTIAL

What is "Maximum Recycling Potential" (MRP)? It is a process through which communities can make an assessment of recycling system needs and, based upon a comparison of costs versus recycling results, decide upon a system ideal for specific local conditions. While the phrase Maximum Recycling Potential sounds simple at first, the full implications of MRP are often not known until significant public resources are invested and results evaluated. The MRP approach to recycling planning will be increasingly important as states and municipalities continue to adjust regulatory goals for recycling. As communities attempt to reach higher recycling rates, the efficiency and effectiveness of specific technical program elements will become critical.

In this section, issues like the makeup of the target waste stream; likely recovery rates from specific technical approaches; and a method for actually calculating an MRP score for a particular community are offered. In Section 2.4, an example from Muscatine County, Iowa of an application of the MRP approach is provided.

2.1 Point of Measurement

One of the most important aspects of using an MRP approach is to understand where the principal point of measurement for determining composition and generation rates for municipal solid waste should be. Much of the planning and studies completed previously in industry have been focused on the point of generation (i.e., residential kitchen, lawn, the fast food dining room, or the computer printer).

While those points and many others provide an indication of gross waste generated, they do not account for what happens to the waste and recyclables that may be diverted as they are transported to processing or disposal facilities. According to Lanier Hickman of SWANA, it is at the gate(s), or the tipping floors of municipal solid waste (MSW) facilities, including transfer stations, waste-to-energy plants, landfills and material recovery facilities where material is accepted for processing or disposition. Emphasis should be placed at those points because it is there that local taxpayers are responsible for facilitating recovery or disposal. Once through the "gates", a specific volume or weight-based charge is applied and routed to municipal coffers.

It would be easy and convenient to identify the point of collection as the appropriate point of measurement, however, consider the following:

- In many communities, responsibility for the collection of residential solid waste is shared between public and private entities. Some communities rely totally on private haulers to service residential units.

- Commercial and industrial waste are generally collected by private haulers with fees paid directly to the private company by its customers with no municipal involvement.

- Municipal governments often lack authority to determine the disposition of commercial or industrial waste. Even with the ability to issue licenses, there is little authority available to determine how the private companies collect or dispose of commercial and industrial waste.

- The majority of existing landfills are owned by municipal government; local governments have partial control of nearly all waste-to-energy facilities; transfer stations and material recovery facilities are owned by local governments.

Part of the theory behind MRP systems is the assumption that system costs and commitments span only that portion of the waste stream for which it has control. If local government cannot or will not control its solid waste stream until it reaches a defined point of disposition, it cannot take responsibility for meeting specific recovery rates from the entire MSW stream and the point of original generation.

2.2 Relating Recycling Goals to MRP

As states and provinces continue to adjust regulatory goals for recycling, it is becoming more and more difficult for local programs to comply. A growing number of states have established recycling goals of 40 percent or more. They have ranged as high as 70 percent (Rhode Island). While often stated as a goal or a guidance policy, many state regulators have placed major emphasis on attaining or making firm commitments to attain the higher goals prior to the approval of permits for other related solid waste management projects.

One typical example of the impact the higher goals can have on municipal solid waste planning can be found in upstate New York. Two counties, working under a joint solid waste management authority, developed a 20 year comprehensive plan. In the plan, assumptions were made that the counties would progress from recycling levels of less than 10 percent in the first few years, to more than 60 percent by the end of the planning period. There was no demonstrated experience on which to support this assumption. This was a particularly dangerous move because the plan precluded the development of sufficient landfill or waste-to-energy capacity to protect local residents should expectations not be met. As a result of underestimating landfill capacity requirements, the counties could expect to prematurely exhaust remaining landfill space leaving residents without disposal capacity while continuing responsibility for long-term debt payments on the prematurely closed facility through the term of the original financing.
To illustrate this problem, let us turn to the existing situation in the State of California.

Under Assembly Bill 939, which was enacted in 1991, communities could face significant monetary penalties for noncompliance with a 50 percent mandatory waste diversion goal by the end of the century. Ironically, experience has indicated that most successful source separation programs nationwide hardly approach a 30 percent recycling level of an entire municipal waste stream. Progress beyond 30 percent limits, particularly without control of the entire waste stream would require significant additional investments in technology. That technology would be required as part of a system to achieve maximum recycling potential.

2.3 Community Specific

Every community has its own MRP. Depending on a range of local conditions, it is possible to calculate MRP. These include:

- The quantity and characteristics of the waste stream; level of control the community has over waste flow;
- Developmental constraints including the availability of suitable sites for specific program elements like composting facilities;
- Environmental regulations, especially those that govern the end use of specific products like municipal waste compost;
- Financial capabilities, materials markets; and
- Public and political acceptance.

The calculation of MRP is highly dependent on an understanding of the MSW stream. The quantities of material designated for recycling and the establishment of reasonable rates of recovery will be important. The impact of related waste reduction programs must also be assessed prior to calculating MRP. Further, a municipal MRP score will not take into effect the efforts of existing or planned private sector recycling programs as they are not under local government control. Thus, accurate performance data may not be available.

2.4 Calculating Your Community's MRP

As you prepare to address your specific community, give careful consideration to the types of material designated for recycling and the availability of sufficient markets to absorb the range of recovered products that you anticipate. While significant amounts of municipal solid waste can be recovered in the form of compost, recovery and recycling are not the same. Before you count a MSW compost product toward your MRP, become familiar with state and local regulatory guidelines that may impact compost marketability. The same advice is true for all other potential products of your recycling system. Your maximum recycling potential equals that material that you can successfully capture, recover, and recycle at a cost that is feasible for your community.

As you review market conditions, you may find that the recovery of a specific product is technically feasible. However, a review of associated costs may reveal that targeting that specific product is prohibitive. For example, in some markets, the cost to refine glass cullet may outweigh recycling advantages. A phased program may be required to introduce more costly methods to recover and market items over time. Under this scenario, you may wish to establish interim goals for your MRP. The Muscatine County, Iowa example illustrates this approach. Keep in mind as you identify final targets that the fewer products recovered will automatically reduce your MRP.

There are three primary steps to calculating your community's MRP:

- Step 1 - Determine the characteristics of the community's waste stream;
- Step 2 - Estimate the likely recovery rates of each recycling program element; and
- Step 3 - Calculate the recycling potential in each and all programs and combine the results.

2.4.1 Step 1 - The Waste Stream

The municipal solid waste consists of several substreams. The composition of the local waste stream is often the source of confusion about recycling results when comparing one community to another. Some communities reference three basic substreams: residential, commercial and industrial. Others classify MSW into more substreams including residential, commercial, industrial, construction & demolition, sludge and institutional waste substreams. Communities may include some or more of these substreams in their definition and records related to MSW management.

Early planning efforts need to include a detailed analysis of the waste stream. If information is not available, a waste stream study can be justified. While many communities rely on national studies to estimate quantities within their jurisdiction, such studies are generally broad based in nature. Local data can differ considerably depending on commercial and industrial makeup and demographics that may present unusual circumstances (for example, a large student population). In order to account for seasonal fluctuations, the impacts of holidays and other variables, a twelve month (or more, if time and resources are available) study will provide an acceptable level of confidence. Data from adjacent communities and those that closely resemble yours, depending upon the quality and protocol used, can be an effective way to cut study costs.

There are two major objectives to conducting the waste stream analysis:

- Determine the composition of the MSW stream targeted; and
- Quantify the materials potentially designated for recycling within the stream.

Three basic methods of waste stream analyses include:

- Waste stream sampling/sorting/weighing;
- Field surveillance; and
- Review of scalehouse data.

Waste sampling, sorting and weighing can be costly and time consuming. Many communities cannot afford a detailed analysis. It involves manual sorting of MSW into representative samples and subsamples. The subsamples could number more than 30 categories. A good example is paper and the numerous subcategories that could be targeted for recovery. As recycling programs are refined and expanded, item specific data like the quantity of junk mail in the waste stream can be used to draw a "roadmap" as to where improvements can be made. Samples must be taken randomly and should not exceed 300 pounds per truck load of waste evaluated. They should be taken during at least two different seasons, if not four during an entire year to account for related seasonal fluctuations.

Field surveillance provides data on the basis of volume for each selected category of waste. It works particularly well for substreams like commercial or industrial waste and construction and demolition debris when the load is homogeneous in nature. The field study team can multiply vehicle volume by density factors to calculate a reasonable weight-based estimate. Table 2-1 provides typical density ranges for different types of loads. Field surveillance can be combined with a review of scalehouse data to allow for estimation of the percentages of the MSW substreams (residential, commercial and industrial). Each substream can be broken into constituent categories by using a sampling, sorting and weighing approach. This combination of techniques will yield the best possible data.

TABLE 2-1

TYPICAL DENSITIES OF MUNICIPAL SOLID WASTE BY SOURCE

		Density (Pound Per Cubic Yard)	
Source	**Compacted**	**Range**	**Typical**
Residential Waste	No	150-300	250
Commercial-Industrial Waste	No	300-600	500
Residential Waste	Yes	500-1000	750

Note: Kilograms Per Cubic Meter = Pounds Per Cubic Yard x 0.5933

Source: SWANA, 1992

When reviewing scalehouse data, weight information is analyzed to determine the relative weights of selected streams as well as to indicate changes in MSW composition on a seasonal basis. This is a useful approach to determine the basic substreams (residential, commercial and industrial). Standard compositional data can then be applied.

For communities that do not have the financial resources needed to fund a comprehensive approach, the use of national or regional data with the assumption that the data represents averages that apply within a community can be helpful. However, there are very few communities that are exactly "average". Under this approach, it is important to look for conditions present in the community that may represent atypical waste streams. For example, some communities in the Carolinas have significant waste generated by furniture manufacturers. Glues, paints, finishes and scrap wood, and textiles will likely present additional waste management challenges for the community. Also, the presence of large institutions like prisons, hospitals, and universities will likely skew community specific data well beyond the "average" results provided by national studies.

As part of determining MRP, it will also be important to identify the quantities of waste managed by local government and private entities. Information reflecting activity within the service area includes waste generation, collection, importation/exportation and final disposal by public and private sectors.

2.4.2 Step 2 - Estimate Likely Recovery Rates

Municipal recycling programs can feature a variety of different technological approaches, each featuring a different potential recovery rate. Those rates will be influenced by the specific feed stream and climate. For example, some communities in Northern New Jersey have significantly more newsprint in their potential feed streams than rural communities in South Carolina. Trade publications like MSW Management Magazine, Biocycle, Waste Age, Solid Waste Technology and World Waste Magazine often publish credible information relative to specific technology oriented recovery rates. Trade associations that deal with a specific commodity will often have reliable recovery rates for a given material from a variety of collection schemes. Examples of those organizations include the Glass Packaging Institute and the Steel Can Recycling Institute.

The major categories of technology that make up the menu of options available to a community include:

- Source separation of materials, commonly known as curbside recycling, involves the separation of certain recyclable materials at the source, either at the residence or business, for later collection, processing and recovery;

- Source separation/composting of yard waste (grass clippings, leaves and brush) requires that residents or businesses keep yard waste separate from the waste stream for separate collection and processing through composting;

- Mixed waste processing involves the processing of either the entire mixed waste stream or select loads of mixed waste for the recovery of recyclables. Mixed waste processing generally involves a combination of mechanical processes and hand-sorting stations at a central facility, which can either stand alone or serve as a "front end system" to a materials conversion process such as composting or waste-to-energy; and

- Mixed waste composting involves processing what remains in the waste stream into compost material for use as a soil conditioner.

2.4.3 Step 3 - Calculate the Recycling Potential of Each Technology

The recycling potential for each technology option can be calculated as follows:

Targeted Waste Stream Fraction x Technology Recovery Rate = Technology Recycling Potential

The "Targeted Waste Stream Fraction" is the portion of the total MSW stream contributed by the specific material designated for recycling. The Targeted Waste Stream Fraction must, in some cases, be adjusted by subtracting the recycling rate of programs instituted "upstream" of the technology component. For example, if local Boy Scouts are consistently collecting aluminum beverage cans, their progress should not be double counted in the community's efforts. This adjustment will reflect activity that diverts material before it reaches the technology component.

The community's MRP equals the sum of recycling potentials of each technology component selected. The MRP may be greater if the community elects to use the complete menu of options. However, economic considerations may preclude comprehensive systems. Listed below are two examples of how to calculate a community's MRP.

Table 2-2 illustrates how to calculate the MRP for a sample community. This example was taken from waste composition studies in Delaware for the Delaware Solid Waste Authority. Given demographic and geographic differences between communities, this information should not be considered to be typical or representative of any specific community other then that which was the subject of the original studies.

Note that this example does not include MSW composting as a technology component. At the point in which the supporting waste composition studies were done, serious doubt existed as to the marketability of MSW compost beyond Delaware borders. Given the importance of product marketability toward achieving MRP for a specific technology, an informed decision was made to not consider this technology due to regulatory constraints. As time passes and more information available relative to the implications of land application of MSW compost, communities may wish to consider this technology as a viable option. Solid assurances of MSW compost marketability would dramatically impact the technology's ability to boost the MRP for a given community.

The second example represents a conceptual level analysis done with generic information for the benefit of a rural county in Iowa. Muscatine County, Iowa was interested in identifying a recycling and waste management system that would help it reach state mandated recycling goals while restricting the cost of the system by using components that would provide the most recycling return for public dollars invested.

Using county-based waste stream data and published information relative to material recovery rates from various technologies, a three-phased system was identified as illustrated in Figures 2-1, 2-2, 2-3, 2-4, and 2-5, respectively. A phasing approach allowed for the extended period of time available for implementing the program and a desire to spread the community's investment requirements over an eight year period. Given the fact that MSW composting was included in the system, it was prudent to delay implementation of that element until regulatory and operational concerns were addressed under Phase III.

Note that the total MRP for Muscatine County under this conceptual plan would be 52 percent of the material controlled by the county. That figure includes the assumption of feasibility of MSW composting as a key component. One element that was not included in this plan was the use of source separation of residential recyclables and curbside collection. Given the relatively low potential of source separation of residential material to reduce the overall waste stream (projected maximum of 6 % under this scenario), the costs associated with the option were viewed as high when compared with the anticipated rate of return. Less expensive, higher return components were selected and programmed for the first two phases of the program. Phase III, which included MSW composting, would require a significant capital investment and long term operations and maintenance costs; however, the potential to recycle 24 percent of the waste stream and the extended period of time (1998 to 2000) until implementation provided justification.

2.5 Other Considerations

Having reviewed the two examples above, you may have noticed the degree to which the commercial and industrial waste streams contribute to a community's MRP. Experience has shown that traditional, source separation/curbside collection programs are not capable by themselves of reaching higher recycling goals. When combined with yard waste composting or mixed waste processing for residential waste, the recycling potential is much greater. Without access to commercial and possibly industrial waste streams within the community, however, recycling rates of 40 percent or more are virtually impossible to achieve. Yet many communities lack the ability to access or control the disposition of commercial waste streams. As mentioned earlier in this section, the community's MRP should be factored downward for that waste that is not controlled by the municipality. Accurate reporting and long term assurances of waste diversion for recycling are a significant problem where there is no flow control. That problem can be compounded by regulatory fines or other forms of penalty associated with an inability to achieve legislated recycling goals.

2.6 Conclusions

Calculating the MRP can help a community set priorities and get a better focus on its recycling efforts. MRP calculations also can provide a preliminary indication of how difficult, and how expensive, it will be to meet mandatory or voluntary recycling goals. Maximum Recycling Potential means just that, the maximum amount of recycling possible given ideal market, regulatory, citizen participation and technological limits.

TABLE 2-2

EXAMPLE OF CALCULATING A COMMUNITY'S MRP

Waste Stream	% of Total Waste (A)	Technology Used	Recovery Rate (B)	Recycling Potential (A x B)
Residential Waste	50%	Source Sep. (SS) Materials	8%	4%
Residential Waste	50%	Source Sep. (SS) Yard Waste Composting	15%	8%
Residential Waste (After SS)	38%	Mixed Waste Processing (MWP)	5%	2%
Subtotal Residential	50%			14%
Commercial Waste	40%	SS - Materials	10%	4%
Commercial Waste	40%	SS - Compost Yard Waste	5%	2%
Commercial Waste (After SS)	34%	MWP	40%	14%
Subtotal Commercial	40%			20%
Industrial Waste	10%	SS - Materials	10%	1%
Industrial Waster (After SS)	9%	MWP	50%	5%
Subtotal Industrial Waste	10%			6%
Maximum Recycling Potential (MRP)				40%

Notes: SS = Source Separation

Source: O'Brien and Williams, 1991

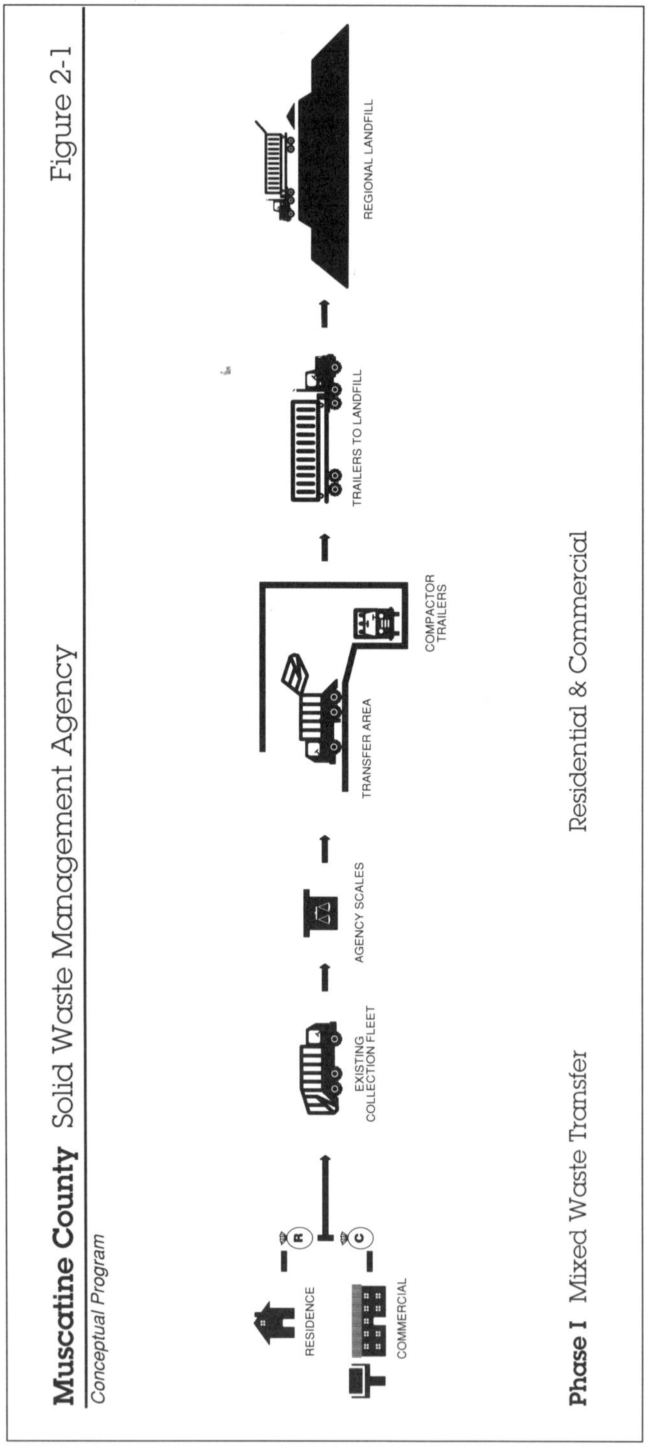

Source: HDR Engineering, Inc., 1991

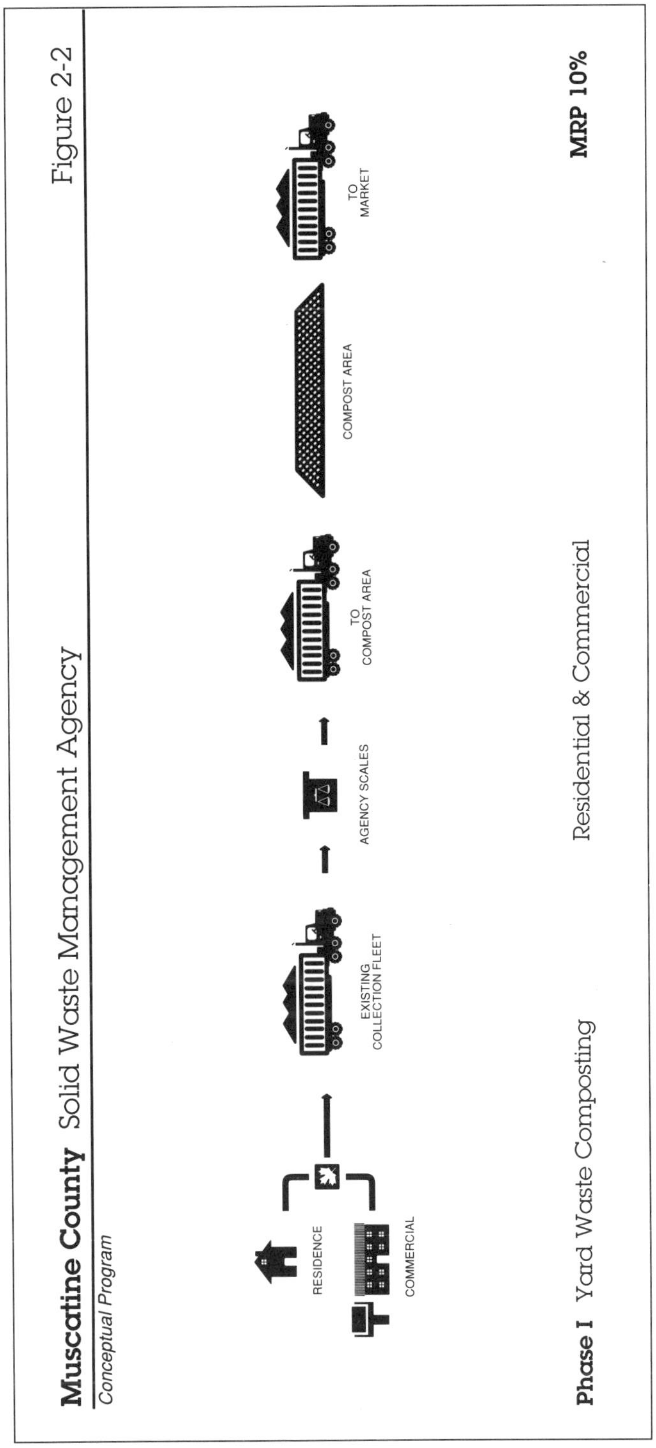

Source: HDR Engineering, Inc., 1991

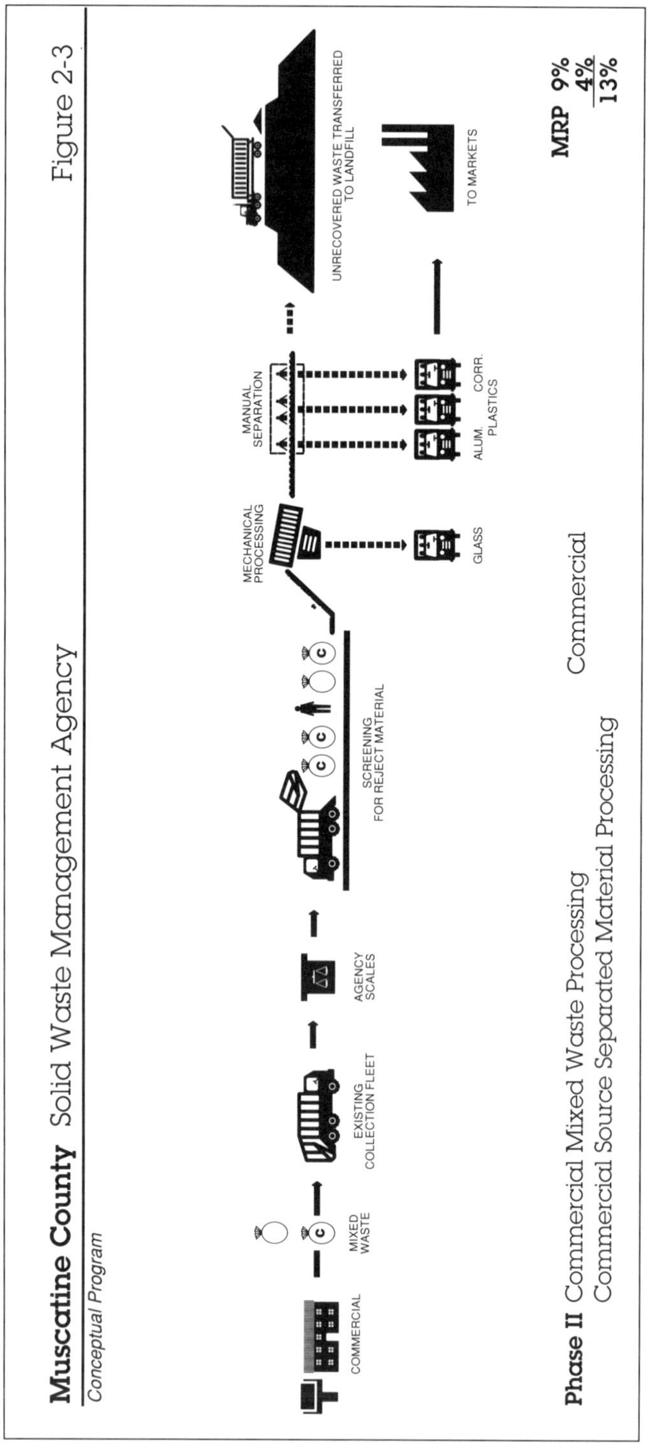

Source: HDR Engineering, Inc., 1991

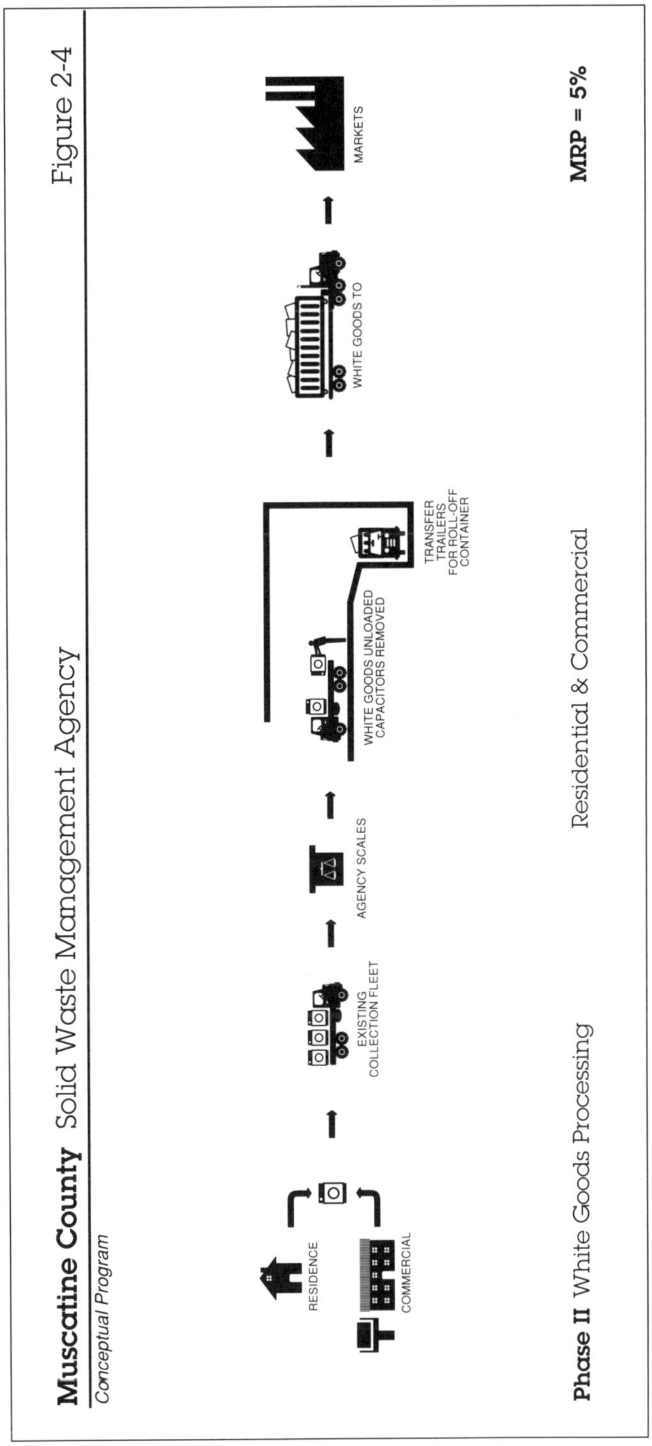

Source: HDR Engineering, Inc., 1991

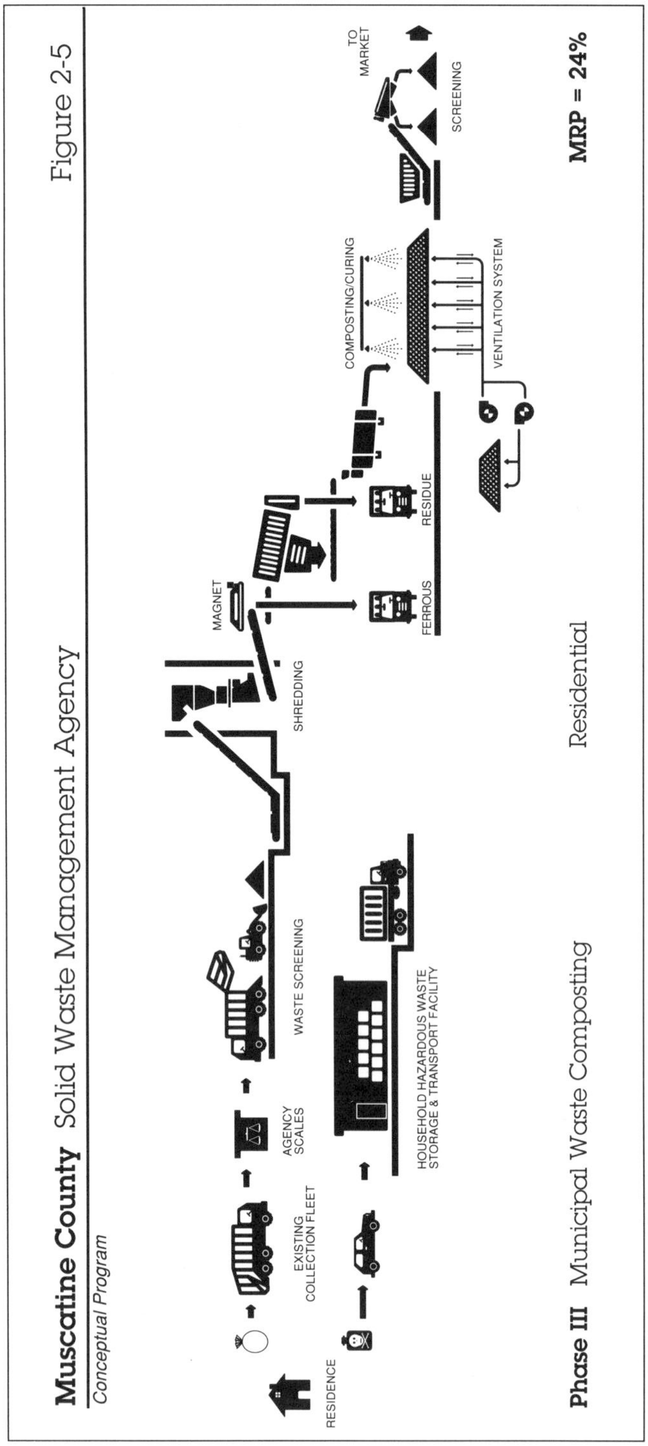

Source: HDR Engineering, Inc., 1991

-3-

COLLECTION APPROACHES

Legislation in many communities requiring both mandatory and voluntary recycling programs has led to an explosion of recycling programs nationwide. There are various options for collection of recyclables, each with its advantages and disadvantages and each with its advocates and detractors. The purpose of this section is to review information on the technologies which are available for recycling materials from the solid waste stream. The concepts discussed are geared towards the collection of recyclables. A discussion of these systems will provide the reader with a general understanding of the types of equipment available and the general advantages and disadvantages. For the purpose of this discussion there are three general types of collection programs used for recycling:

1. Curbside Collection
2. Permanent Collection Centers; and
3. Periodic Collection

Moving from one to three, the programs decrease in cost, complexity, and the need for staff labor. The programs also decrease in convenience for the public, generally resulting in less public participation.

3.1 Curbside Collection Alternatives

Within most communities, solid waste from single-family residences, multi-family units, and businesses are typically collected at the curb or alley in various-sized plastic and metal containers or plastic bags. Generally, wastes are collected once a week for multi-family units (apartments and condominiums) and several times per week for commercial establishments depending on their service requirements. In contrast to residential recycling programs which produce generally uniform waste streams, a commercial recycling program must contend with unique waste streams produced by diverse generators.

Prior to significant government involvement in recycling, residents and businesses wishing to recycle materials had to locate a buyer for their recyclables, transport the materials to a buyback center, or rely on periodic collections made by volunteer groups. Today, counties and municipalities are encouraging greater recycling participation by making it easier for residents and businesses to recycle by providing curbside collection of recycled materials. Once the materials are collected, the county, city, or private entity (e.g., hauler) takes on the responsibility of processing them so that they can be marketed. Participation in these programs range from a few percent to 80 to 90 percent, while most programs achieve 40 to 60 percent participation (Table 3-1).

TABLE 3-1

COMPARISON OF RECYCLING PROGRAMS TONS RECOVERED AND PARTICIPATION RATES

City	Total Pop.	Population Served	Househol d Served	Tons[1] Recycled	Collection Day Participation Rate	Overall Participation Rate	lb/[2] Capital/ Year	lb/[3] Household/ Year	Percent of Recovered Material[5]		
									Newspaper	Glass	Aluminum/Tin
Austin, TX	450,000	225,000	90,000	5,400	10-12	20-25	48	120	78	19	3
Davis, CA	47,000	NA	11,000	3,200	50	60	218	581	66	16	3
Evesham, NJ	36,000	30,000	8,500	2,995	50	85-90	199	704	65	25[4]	-
Groton, CT	10,000	NA	1,900	626	50	75-85	247	658	65	35[4]	-
Haddonfield, NJ	12,500	NA	4,400	1,5000	66	95	256	681	67	33[4]	-
Marin Co., CA	117,000	117,000	44,000	12,500	35-40	60	213	568	68	30	2
Mississauga, Ont.	400,000	301,000	90,000	14,000	40	80	93	311	70	15	10
Niagara Falls, Ont.	70,000	60,000	19,500	2,307	45	75-80	77	236	70	21	9
Plymouth MN	43,000	NA	12,500	2,800	NA	53-56	168	448	75	19	6
San Jose, CA	720,000	540,000	180,000	24,000	25	58	89	266	70	27	3
Township, PA	22,000	22,000	6,800	1,972	60	70	70	580	63	22	-
Sunnyvale, CA	116,000	NA	28,000	4,078	21	50-60	109	291	73		5

NOTES:
(1) Includes newspaper, glass aluminum and tin.
(2) Based on total tons recycled/population served; 2.66 persons/household assumed where information is not available.
(3) Based on total tons recycled/households served.
(4) Includes glass, aluminum and tin.
(5) Percentage of program recycled tonnage.
(6) NA denotes information not available.

SOURCE: BioCycle, 1992.

Collection involves three basic steps: (i) source separation, (ii) collection in one of a variety of collection vehicles, and (iii) hauling to a processing destination or final market. The waste generator typically has some responsibilities for separating the recyclable materials from the waste stream. The type of collection fleet used and the design of the facility for processing recyclables separated from the waste stream are the major factors in determining how much source separation is necessary. Collection costs can be reduced by siting processing facilities near the centroid of material generation or by providing transfer stations. Various collection alternatives available for recyclable materials from multi-family units and commercial businesses are described below. In general, such programs require residents to separate recyclables from other household refuse and place them at the curb for a regularly scheduled pick up. As shown in Table 3-2, depending upon the level of processing available, curbside programs may require complete separation, three-bin containers, single-bins or plastic bags, where materials such as glass and aluminum are allowed to be mixed. Each curbside collection system has its own inherent advantages and disadvantages based on recent experiences by many curbside collection programs.

3.1.1 Source Separation

The first alternative generally used for multi-family and commercial units is a source separation program, although the physical structure and management of these type of complexes requires that different approaches to recyclables collection. For example, a densely populated, high-rise unit has different needs than a two-story condominium or garden apartment and often has different forms of solid waste removal programs, service styles, locations, and type of collection vehicle.

Generally, however, residents and businesses are most often asked to keep a number of recyclables separated from the mixed waste by storing them in centrally located bins or metal dumpsters. The residents or businesses then subsequently place these bins at the curb for pickup. This requires, however, that building management must provide storage space for recyclable materials which is typically at a premium at these locations. In high-rise, multi-family units, for example, recyclable materials may be stored on each floor. However, in New York City's multi-family program, each building is given a two-cubic yard, front loading container which is emptied weekly. The materials may or may not require subsequent processing to meet market specifications.

TABLE 3-2

ADVANTAGES AND DISADVANTAGES OF CURBSIDE COLLECTION SYSTEMS

Material Collection System	Advantages	Disadvantages
Complete Separation	Easy to implement by community	Poor citizen participation
	Little or no processing required	Major public information program required Multiple Containers
Three-Bin System	Provides peer pressure effect	Expensive to implement
	Convenience to homeowner	Increased processing costs
	Higher citizen participation	May require specialized collection equipment
One Bin System or Plastic Bag	Most convenient to homeowner Higher citizen participation Relatively inexpensive	Increased processing costs

Information is critical in these programs since the often single and mobile lifestyle of many multi-family tenants makes communication difficult. There is also high turnover in multi-family units. Added to these communication difficulties is the potential that residents in lower income units often do not speak or read English as a first language. Consequently, information programs should be designed to recognize this need. Given the problems in establishing these programs, many cities have been reluctant to expand their community-wide recycling collection programs to include multi-family recycling.

High material quality of recyclables is generally achieved through source separations. Specialized collection trucks must be used. However, this collection approach is labor intensive. Participation is generally lower than is the case with other approaches because more effort and space are required to keep the materials separate. This concern increases with the number of materials recycled.

The separate collection of yard waste also poses unique problems based on the type of collection equipment which is available in the community, the composition of materials deposited curbside (leaves, grass, or brush), seasonality and frequency of curbside pickups. In many communities in northern climates, for example, yard waste and leaves are raked into the street in the fall when it is commonly collected using mechanical claw or vacuum leaf collector trucks. Front-end loaders can also be equipped with scoop or pincer devices, force-feed loaders, or vacuum machines which can be used in connection with another unit such as a dump or packer truck. To reduce the time needed for collection, restrictions are often required as to the size of the piles, location of piles, or restrictions for parking vehicles on streets. Unfortunately, loose yard waste collection has disadvantages in that more extraneous material ends up in the waste, leaves, and clippings which create problems in the street (safety, aesthetics, and stormwater concerns) and it does not work well in wet weather. Specialized collection equipment, which may be needed in this case, has associated costs and street sweeping may be needed.

In other communities, standard plastic garbage cans or wheeled 60/90 gallon containers are used with these materials being traditionally collected on specific days of the week on separate yard waste collection routes. The advantage of using containers or bags is that they minimize blowing leaves and grass, are not sensitive to weather, and they keep out contaminants. The use of these containers eliminates the necessity for manual debagging of the yard waste, but results in high initial start-up costs, especially if the community must purchase the large rolling containers and modify their existing collection fleet with "flippers" since a 60 or 90 gallon container is quite heavy to lift. One concern for collection systems using rigid containers is that some programs have experienced problems with participants putting other types of waste in the containers and contaminating the yard waste. The best way to avoid this problem is to educate the users as to the problems created and the environmental concerns. Well educated communities that allow their citizens to "buy in" and support the program are less likely to have this problem.

Generally, most communities require their residents to utilize standard plastic garbage bags, or biodegradable kraft paper bags to minimize capital costs. Plastic bags are widely available at most stores and minimize disruption to homeowners. Clear bags allow collection crews to examine the contents for contaminants and instructions can also be printed on the bags. However, plastic trash bags interfere with composting or mulching equipment and do not biodegrade. Thus, the community often has to decide whether to reduce its collection efficiency and debag during curbside pickups, or as an alternative, install mechanical bag separation equipment or additional employees at the composting facility to manually remove the plastic bags prior to introduction of the yard waste into a tub grinder. Removal of the plastic bags is essential, especially if an end market is sought for the compost product.

Biodegradable bags are another option since they may save debagging time at the curb or the composting facility. However, it should be noted that some existing community programs have had problems with biodegradable bags in that they did not break down as expected.

3.1.2 Source Separation, Curbsort

The curbsort alternative requires residents to place all recyclables into a single bin and at the bin at the curb. The recyclables are then sorted at the curb, by the collector, into the separate compartments of the collection truck. Both material quality and participation rates are reportedly high using this approach because a single bin is convenient and the curbside sorters perform a quality control function by rejecting non-recyclables. Some studies have shown that curbside sorting can be extremely efficient. However, curbsorting is much more labor intensive at each collection site and there is an upper limit to the number of homes that can be economically collected as the size of the route increases and the number of materials is collected. Generally, this means that the curbsort approach is applicable only to the single-family, residential waste stream. However, small and medium-sized communities which lack capital funds for processing facilities should carefully examine this option.

3.1.3 Source Separation, Commingled

The commingled alternative required residents to place recyclables into a single bin, which is placed by the resident at the curb. The materials are then collected using a single- or dual-compartment truck, with newspaper often kept separate. The benefit here is greater utilization of truck capacity with only two compartments since it is much easier to relocate the divider to maximize available truck capacity. The commingled recyclables are then subsequently separated at a Material Recovery Facility (MRF), adding another expense to the total recycling program budget. When commingled recyclables are placed in a collection vehicle, collection time is saved resulting in extending route sizes, but processing time is increased at the processing center because the collected materials must be separated, cleaned, baled, shredded, or otherwise prepared according to market specifications.

3.1.4 Source Separation, Combined Collection

With the combined collection approach, residents or businesses set out recyclables in commingled form, in a separate, easily identifiable blue or other translucent bag. The mixed waste (newspapers, glass containers, HDPE, PET, steel/bi-metal cans, and aluminum cans), are then collected in existing packer trucks, either alone or together with the mixed waste, depending on the program. Collected materials are delivered to a facility for separation as required. Recyclables are directed to a processing facility and waste is directed to the existing disposal facility. This combined collection approach has grown out of a need by communities to provide more cost-effective curbside recycling programs than using the traditional approaches.

The collection of commingled recyclables in compacted form is being tested in a number of jurisdictions (Table 3-3) using blue plastic bags, typically 13-gallon (1.01, 1.5, or 1.75 millimeter thick). In this approach, regular refuse packer trucks are used to collect the commingled recyclables, using some degree of standard compaction, with care taken not to incur excessive glass breakage. There are two approaches in the adoption of this alternative: addition of a separate collection service or modification of the regular collection service to accommodate recycling.

In the first approach, a jurisdiction may use existing packer trucks to collect recyclables through the addition of a recyclables collection service. This can be done by collecting the recyclables on collection off-days (generally Wednesdays for jurisdictions providing twice/week collection). The addition of a recyclables collection service results in additional collection costs. However, as indicated above, no additional capital costs are incurred for collection equipment. To date, there has been little research to compare the efficiencies of the bin versus the bag type approaches, although the blue bag type program has been criticized by its detractors because of purported increases in glass breakage and material contamination of the contents of the blue bags as more materials are collected at higher compaction rates. There is also some concern that participation in such programs may suffer when residents must pay for the bags or pick up the bags at specific locations.

Minnesota's South St. Louis County Solid Waste Commission also offers a variation of the combined collection approach to all 3,000 households in its 1,180 square mile service area. Residents put recyclables into blue recycling bags, provided by the Commission, which are placed at the curb for collection with the mixed waste. The blue bags are separated from the mixed waste at a transfer station and are subsequently transported to a MRF for processing and sale. Net costs for this rural program, including bag costs and sorting at the transfer station, are reported to be $ 0.89 per household per month. The Commission estimates that to have offered traditional curbside pickup to its geographically spread-out residents, the county would have needed to purchase two or three new trucks, raising the project costs for curbside collection to approximately $11 per household per month.

The second approach, the modification of the regular waste collection service to incorporate recycling, offers the most substantial cost savings of any collection alternative. In this approach, the collection service is generally changed from twice per week mixed waste collection to once per week mixed waste collection and once per week mixed recyclables collection. In this approach, no additional collection costs are incurred. Since collection costs typically represent 50-75 percent of the costs of a curbside recycling program, these cost savings are reported to be substantial.

3.2 Permanent Collection Centers

Although in the last decade, many communities have launched extensive curbside recycling programs, there has been a trend in recent years to back away from curbside recyclables collection due to its high costs and the low prices paid for recyclables. As a substitute, these communities have replaced their curbside collection programs with the establishment of permanent collection centers. Permanent collection centers are usually referred to as recycling or dropoff-centers. These centers require that residents bring their recyclable materials to the site and place the burden of separation on the residents. Oftentimes, residents are asked to bag or tie newspaper, to separate glass by color, and rinse out all containers.

Permanent drop-off centers can vary greatly and are commonly employed in recycling programs. Though drop-off centers typically divert only a reported 1 to 10 percent of the residential waste stream, these type of centers serve as a good starting point for a community to use in implementing a comprehensive recycling program and can serve as a backup to a curbside collection program in case homeowners miss a pickup or need to remove materials prior to their normal pickup day. They are also an excellent way to provide residents in multi-family housing and businesses to participate in a recycling program even if they are not currently serviced by a curbside program. Also, the drop-off programs allow a community to experiment with the collection of new materials prior to inclusion in its curbside program.

Drop-off centers can be designed to collect a broad range of materials, including aluminum, glass, newspaper, old corrugated containers, and even high-grade office paper. In order to reduce costs, drop-off centers are often located on free property, such as municipal lots, schools, church property, and sometime shopping center parking lots. They can be as simple as a few collection bins or as complex as a large processing center equipped with state-of-the-art technology located at disposal sites. Drop-off centers may be open to the public 24 hours a day or have established hours of operation. Some centers are staffed to provide assistance to recyclers in unloading vehicles, processing any mixed materials, and keeping the centers neat and clean. Other centers may be unattended, except for pickups of collected materials. Instructions regarding material deposit at unattended centers are usually highly visible and clearly understandable. Regular visits to the site by staff are usually scheduled to ensure that areas are neat and clean and containers are not overflowing with materials. Security considerations should dictate that these areas are well lighted and are close to nearby roads and other public facilities to avoid vandalism or illegal dumping. The attended sites, while more costly, avoid these potential security problems.

The more complex drop-off centers may provide some types of material processing for volume reduction or improvement of material marketability. These complex centers, therefore, require additional equipment such as sorting tables, balers, conveyors, shredders, magnetic separators, can flatteners, glass crushers, and fork-lift trucks in addition to the normally required weigh scales and storage containers. Staffing must be commensurate with the operational requirements of such a facility.

Factors affecting the quantity of materials recovered under a drop-off center program include the following:

- Program promotion;
- Local waste composition;
- Types of materials collected;
- Number of centers in a given geographic area;
- Site maintenance;
- Site security; and
- Access to the centers.

TABLE 3-3

SURVEY OF BLUE BAG RESIDENTIAL RECYCLING PROGRAMS

Location	# of Households Served	Newsprint	Corrugated Cardboard	Ferrous Cans	Alum.	HDPE	PET	Glass
Co-Collection Systems								
Houston, TX[1]	18,300	X	No	X	X	X	X	No
St. Louis County, MN	2,800	X	X	X	X	X	X	X
Omaha, NE	100,000	X[2]	N/A	X	X	X	X	X
Missoula, MT	17,000	X	X	X	X	X	X	No
Whitman County, WA (Pullman)	26,000	X	X	X	X	X	X	X
High Point, NC	27,000	No	No	X	X	X	X	X[4]
Baltimore, MD	233,000	X[2]	N/A	X	N/A	X	X	X[6]
Separate Collection Systems								
Pittsburgh, PA	170,000	X[2]	N/A	X	X	X	X	X
Danbury, CO	20,000	X[2]	N/A	X	N/A	X	X[5]	X
Ft. Myers, FL	9,180	X[2]	N/A	X	X	X	X	X
Pasco County, FL	90,000	No	N/A	X	X	X	X	X
Lakewood, OH	20,000	X[2]	N/A	X	X	X	X	X
Londonderry, NH	6,500	X[2]	X	X	X	X	X	X
Lubbock, TX	5,500	X	N/A	X	X	X	X	X

TABLE 3-3 (continued)

SURVEY OF BLUE BAG RESIDENTIAL RECYCLING PROGRAMS

			Recovery		
Location	Hauler	Collection Frequency (No. of times per week blue bags picked up)	Time Period	Total Tons Recovered	Residue Rate (%)
Co-Collection Systems					
Houston, TX	Private	1	11 months	811	N/A
St. Louis County, MN	Private	1	N/A	N/A	N/A
Omaha, NE	Private	1	12 months	N/A	15-20
Missoula, MT	Private	1	13 months	N/A	15
Whitman County, WA (Pullman)	Private	1	N/A	N/A	10
High Point, NC	City	2	N/A	N/A	N/A
Baltimore, MD	City	1	3 months	3,506	N/A
Separate Collection Systems					
Pittsburgh, PA	City	1	8 months	5,700	10
Danbury, CO	Private	1	N/A	N/A	N/A
Ft. Myers, FL	Private	2	12 months	797	6.3
Pasco County, FL	Private Haulers	1	N/A	N/A	20
Lakewood, OH	City	1	12 months	6,000	N/A
Londonderry, NH	Private	1	18 months	N/A	Low teens
Lubbock, TX	City	1	21 months	82	N/A

NOTES:
(1) This was a demonstration program conducted by BFI for the City of Houston.
(2) Newspaper collected separately.
(3) Diversion rate attributable to blue bag program 3 percent. Remainder is yard waste.
(4) Green glass not accepted.
(5) Milk jugs only. New program will add additional plastics products as participants.
(6) Soda bottles only. New program will add additional become more educated in preparing materials properly. glass products as participants become more educated in preparing materials properly.

SOURCE: HDR Engineering, Inc.

Table 3-4 presents the quantities of materials collected under a variety of drop-off programs in the United States. As shown, the amount of material recovered ranges between 7 and 53 pounds per capita per year. Further, most communities see participation rates in the range of 5 to 20 percent. Although publicity and promotional campaigns normally increase public awareness and participation levels, facility convenience, cleanliness, accessibility, operational hours, and market preparation requirements are the major factors that affect citizen participation levels.

Permanent collection centers can be effective tools in collecting recyclables, especially in rural or low population areas where curbside collection or buy-back centers may not be cost-effective. Drop-off centers are also used in urban and suburban areas to serve multi-family residences and less densely populated single-family neighborhoods where other recycling approaches may not be appropriate. Such centers can include a variety of different kinds of public or private buy-back centers, where generally money is paid for the recyclables, as well as unmanned sites where materials are collected for free. These centers can also be used to collect special wastes such as automobile or household batteries, used oil, and household hazardous wastes. Communities which have developed extensive networks of such centers have spent considerable efforts in siting centers in convenient locations, as well as designing centers that can be efficiently maintained and serviced.

3.3 Buy-Back Centers

Buy-back centers incorporate features of the drop-off center, and are targeted towards residential waste by being located near to residential areas. Yet, they are also similar to recycling businesses such as waste paper and scrap meter dealers in that they provide economic motivation by paying cash for recyclables. Since buy-back centers offer this economic incentive, citizens may be willing to travel further to a buy-back center than a drop-off center. Hence, buy-back centers usually have a larger service area than drop-off centers.

The types of buy-back centers can vary dramatically. The more complex centers will purchase a wide variety if materials and often conduct materials processing such as glass crushing, densifying aluminum, and baling waste paper. Therefore, buy-back centers may require sorting tables, conveyors, balers, shredders, magnetic separators, can flatteners, glass crushers, and fork-lift trucks in addition to weigh scales and storage containers. Several employees may be needed to operate such a facility.

Participation by individuals is based on factors similar to the drop-off center: convenience in location, hours of operation, and materials preparation requirements. Locating a buy-back center on a major thoroughfare allows visibility and easy access increases participation. Promotion and advertising also influences participation, and the effectiveness of the buy-back center depends on a more constant level of advertising.

TABLE 3-4

DROP-OFF PROGRAM CHARACTERISTICS

Location	Population Served	# of Sites	Population/ Site	Tons/yr	Residential Recovery (1) Rate (%)
Arcata, CA	12,340	2	6,170	500	5.0
Champaign County, IL	171,000	15	11,400	1,000	1.5
Columbia County, PA	50,000	17	2,900	469	2.5
Cook and Lake County, IL	270,000	18	15,000	7,140	4.5
Delaware County, PA (2)	500,000	50	10,000	1,800	0.7
Durham County, NC	120,000	10	12,000	1,200	2.5
Fairfax County, VA	75,000	8	9,400	1,000	3.5
Fairview, OK	3,200	2	1,600	66	2.0
Kent/Ottawa County, MI	650,000	30	21,600	3,200	0.9
Mecklenburg County, NC	511,000	18	28,389	3,000	0.9
Morrison County, MN	29,311	18	16,283	722	6.0
Owasso, OK	12,000	3	4,000	372	2.0
Peterborough, NH	5,000	1	5,000	546	18.0
Prairie duSac, WI	2,145	1	2,145	288	21.0
Santa Monica, CA	70,000	66	1,000	1,398	5.0
Snohomish County, WA	NA	15	----	233	2.5
State of Delaware	667,000	100	6,670	7,757	2.0
Tampa, FL	285,000	15	19,000	3,581	1.5
West Greenwich, RI	3,500	1	3,500	216	9.0
Windham, VT	22,000	1	22,000	500	4.0

Notes:

(1) Residential recovery rates of residential waste stream estimated by dividing tons per year by the assumed residential waste generated: 1.75 to 2.5 lb/capita/day for smaller population served; 2.5 to 3.3 lb/capita/day for greater population served.

(2) Collects glass only.

Sources: BioCycle, 1989
Resource Recycling, 1992
HDR Engineering, 1992

In general, establishing buy-back centers is more costly than drop-off centers because of the additional equipment and labor needs. However, a simple buy-back center with very limited processing can be started with scales and storage areas or containers. As with drop-off centers, use of free land and donation of equipment can reduce costs. Since cash is paid for materials, an additional expense factor is a ready cash reserve to purchase materials from the public until a sufficient quantity is accumulated for sale to the buyer.

A number of recyclers and end purchasers have attempted to reduce the costs of constructing and operating buy-back centers by installing automated facilities using reverse type vending machines (RVs) in convenient locations to accept and process beverage containers such as aluminum, steel, glass, and plastics. In most cases, customers receive coin payments based on the weight of materials deposited and current value of the materials. Some operators have also provided coupons and prizes to encourage the level of public participation. It is estimated that RVs currently account for one-fifth of the 50 percent of the aluminum recovery rate.

RV machines have often been touted as a means to reduce high labor costs. However, manufacturers of these machines have seen relatively little growth in sales in the past five years. Many recyclers have shied away from purchasing these machines because of their high initial purchase costs ($5,000 to $25,000) and service requirements. This situation may change as manufacturers begin to offer lower cost models. In recent years, manufacturers of RV machines also appear to be making major strides in developing models which are more consumer-friendly and can accept a variety of materials. The largest user of RV machines has been the aluminum beverage can industry as a means to boost can recovery. Surveys conducted by the Reynolds Aluminum Recycling Company have suggested that RV machines tap a sector not traditionally provided by buy-back centers. These customers are usually small families which are attracted by the RVs convenience and savings in time.

3.4 Periodic Collection Programs

Periodic collection programs require the public to separate out of the waste stream and temporarily store certain recyclable items, which must then be delivered to a specific location at a particular time. Generally, collection vehicles are only provided at the specified location on designated collection days. An example of a periodic collection program is a newspaper drive or aluminum can drive. Volunteer civic groups and churches often operate periodic collection programs, but municipalities sometimes sponsor these programs as well.

3.5 Other Solid Waste Collection Programs

3.5.1 Office Paper Recycling Programs

Office paper recycling programs are designed to separate and recover recyclable waste paper prior to disposal into the municipal solid waste stream. Surveys have shown that general office operations generate approximately one-half pound of recyclable waste paper per employee per day. In order to recover a majority of this material, the following information must be identified:

- Participating areas such as suites, floors, and buildings;
- Number of employees housed in each participating area; and
- Determination of existing disposal practices.

In cases where businesses are spaced out among satellite offices, suites or buildings, the areas with the highest employee populations, and therefore the highest waste paper generators, are generally chosen to initiate an office paper recycling program. Once the program has been established, other satellite offices and buildings can be incorporated into the program expansion. This expansion can be accomplished by allowing other offices in the same building or other suites on the same campus to piggy back onto the existing collection contract in place at the time.

Convenience to the individual employee is the key factor in designing a successful office paper recycling program. The number of employees per building or office determines the quantity of recyclable waste generated and the number of individual and main storage containers required to provide adequate service to the employee. Finally, detailed layout of the internal organization of the program is accomplished by studying the floor plan broken down by floor level and department. Analysis of daily operations within this structure will indicate the most efficient locations of collection containers.

Separating recyclable waste paper from an employee's desk is only the first step in capturing this material from the waste stream. Investigation of existing disposal procedures by the janitorial staff will dictate the collection procedure and location of main storage containers. This is the most important step, otherwise the separated paper may still end up in the waste stream or, worse, create a little nuisance around the office. In order to ensure the success of the program, the separate collection of separated recyclable waste paper by the janitorial staff must become a time-saving step for them rather than an extra burden.

Within the completion of the program technical structure, attention must be concentrated in program management, coordination, and employee education and awareness. This can be accomplished by either appointing an existing employee or having a person to coordinate the extra effort. Additionally, people should be designated as department, floor, or building recycling liaisons in order to answer employee questions or problems. Employee awareness can include seminars with video and slide presentation, recycling posters throughout the office or building, periodic memos answering questions or stating program effectiveness, and a graphical presentation indicating current recycling rate or status towards a predetermined goal. Although these components are important in promoting the recycling concept to employees, they are by no means the standard. Every education program for office paper recycling is only limited by the energy and imagination of the individuals managing the program.

3.5.2 Construction and Demolition Debris Recycling

In addition to residential and commercial solid waste, there is a certain amount of construction and demolition (C & D) debris present in a community's waste stream. This portion of the waste stream comprises bulky, difficult to process material such as wood, tires, concrete, rugs and carpets, large pieces of metal, corrugated containers, wire and strapping metal and the rock and dirt. Generally, C & D haulers will perform limited processing in order to recover the larger pieces of waste metals and, perhaps, concrete and corrugated cardboard for which they have ready buyers.

The major goal of processing C & D for recovery of materials is to reduce the volume of C & D waste to be collected, transported, and disposed. Some vendors claim a nine to one reduction can be obtained through processing. A typical processing scheme consists of initial screening with crushing of the oversized material. The "unders" are then hand picked for recoverable material and the remainder shredded and/or composted. Typical equipment includes grapples, trommels, vibrating screens, hammermills, balers, and dust control systems.

3.5.3 Industrial/Commercial Programs

The solid waste generated by commercial and industrial businesses differs from the residential waste stream. In general, corrugated cardboard, computer paper, and office paper are found in significantly greater concentration in the commercial waste stream. To some extent, waste stream composition is business or industry-specific. Obviously, waste from a restaurant is not like that from a newspaper publisher. The following discussion highlights types of commercial and industrial recycling collection programs that can target specific waste materials.

Commercial and industrial recycling collection programs usually are not operated by governmental entities. The programs are best administered by the business involved. The function of government is best filled by serving as an information resource, education business representative, and providing assistance or guidance with program development. There are likely a number of businesses in your community that are currently involved in recycling. Bank, insurance, telephone, and other utility and service companies often have portions of their paper waste delivered to a paper broker and shredded for confidential reasons. This type of solid waste, in essence, never becomes part of the municipal waste stream.

Office paper recovery programs can be implemented in a wide range of businesses, especially highly computerized ones which use a lot of high-grade paper. Such a program could be expected to recover five pounds of wire ledger paper per employee per month. Bars and restaurants can be targeted with glass and beverage container recovery programs. Recovery of corrugated can focus on grocery stores, warehousing, and supply businesses. A number of grocery stores are probably involved in corrugated recovery in your area.

Viewing waste material as an asset and indirectly providing a savings in waste hauling costs is the basis for implementing commercial/industrial recycling programs. To be implemented successfully, such programs must demonstrate corporate support and cooperation, economic viability, sufficient amount of quality material, and in-house adaptability. Often, facility managers or building service departments are involved in establishing these programs. In-house collection procedures, labor requirements, storage facilities, employee educational programs, materials purchase contracts, and performance evaluations are program elements that must be carefully addressed.

3.5.4 Industrial/Commercial Waste Exchanges

Industrial waste is also likely to be unique to the type of industry and is probably comprised both of solid and liquid material. An industrial waste exchange is a process whereby the by-products or waste products of a particular industrial process are either reused by that industry or another industry. Waste exchange is resource recovery at the industrial level through recycling and reusing valuable materials instead of disposing them.

Two types of waste exchanges exist: passive (information) and active (material). A passive waste exchange is an entity that acts as an information clearinghouse. The former is generally operated by a non-profit association, trade group, government agency, or university. The latter type of exchange is typically operated by a for-profit business. Through a catalog or brochure, generators list waste they have available and want to transfer and potential uses which may be compatible. A passive waste exchange is a matching service. The exchange then contacts the listed party, telling the firm the name of the interested party. At that point, it is up to the listed firm to contact the respondent. The clearinghouse has completed its job at this point. An active waste exchange is an entity that becomes physically involved in the handling of materials.

-4-

PROCESSING TECHNOLOGIES

When considering waste or recyclables processing technologies, it is important to give careful thought to both short and long term objectives. These objectives will be strongly influenced by state and local recycling requirements, particularly recycling goals. Most states have taken different approaches relative to specific percentage requirements for various materials within the waste stream. States like Connecticut clearly ban a long list of specific recyclables from land disposal or combustion. Other states like Pennsylvania give broader discretion to local programs by specifying percentage goals and a minimum number of items to be recycled leaving final decisions as to exact municipal program elements to local planners. New Jersey, which has a 60 percent recycling goal by 1995, established percentage recycling goals for a number of new materials and waste streams not traditionally classified as municipal solid waste (i.e., construction and demolition waste, junked automobiles and industrial scrap).

As you plan to implement a new recycling program or expand an existing one it is important to begin by establishing current per capita waste generation and material recovery and recycling rates. With this information in hand, you can begin to match the technological options available, with actual program needs in order to reach or surpass regulatory requirements and program goals.

This section includes detailed discussion relative to the choices available to recycling coordinators from technological and collection interface points of view. A range of processing alternatives will be discussed including mixed waste processing, municipal waste composting, yard waste composting and construction and demolition waste recycling. The section will be concluded with guidance as to alternative evaluation.

4.1 Recycling Program Alternatives - Processing

In this section detailed descriptions are provided for the following alternatives:

- Mixed waste processing, both low and high technology approaches;
- Yard waste composting, including the composting of leaves, grass and chipped brush as well as the chipping of woody branches to be used as mulch;
- Mixed waste composting, including aerobic (oxygen rich) and anaerobic (lack of oxygen) approaches; and
- Construction and demolition waste recycling.

4.2 Mixed Waste Processing

Studies have shown that, even after intensive residential recycling, large quantities of recyclables remain in the waste stream. Mixed waste processing is usually employed as second chance to recover recyclables from the mixed waste stream that have not been captured by other source separation efforts. They are often developed as part of some other waste management approach like composting or waste-to-energy.

They can be either high (mechanically oriented) or low technology (hand sorting) or combinations of both. It is generally capital intensive and yields relatively low recovery rates (assuming that recycling has been attempted prior to mixed waste processing). The approach has been used in many locations and is closely related to technology used to produce refuse derived fuel.

This approach can be used in conjunction with mixed waste collection and is compatible with residential, institutional, commercial or industrial waste streams.

4.2.1 Low-Technology Material Recovery Systems

Using a low-technology approach, the capital cost is minimized and the recovery process is usually labor intensive. This type of approach is typical of private scrap/recycling operations developed in the U.S. during the 1980's and early 1990's. It enables the system operators to more readily respond to volatile fluctuations in demand and material prices. If a material becomes uneconomical to recover, efforts to recover it can be ceased fairly easily by reducing staff, and start-up costs are not high when markets rebound.

This type of system relies heavily on handpicking, with some assistance from mechanical systems. A typical large scale system would use a tipping floor to receive waste. Pickers on the floor would segregate large items such as white goods or corrugated paper to reduce burden depth and handling problems on the processing lines. This would also enable the recovery of corrugated paper before contamination degrades the quality and marketability of the product. A loader would feed conveyor lines with the remaining waste stream. Further, handpicking of materials (such as cardboard) could occur off this conveyor. A bag breaker would most likely be required for high-volume plants to allow for adequate inspection of the waste stream. A screening process could also be used to separate the presized material (normally five inches or less) from the oversized material and to decrease the burden depth on the conveyor.

After separating material into different size streams, both would be subjected to magnetic separation for the removal of ferrous materials. The undersized fraction would contain most of the cans and bi-metal products. The oversized fraction would contain most of the heavy and more valuable ferrous products. Conveyors would carry both streams past picking stations where the materials would be separated based upon current market conditions. The oversized fraction would contain film plastic and PET containers, as well as most of the paper, cloth and wood. The undersized fraction would contain most of the aluminum cans, glass containers, yard waste, dirt and grit, and food wastes.

The separated material would be dropped down chutes to be processed as needed to meet material market specifications. This processing might include shredding, baling, screening, washing, or magnetic cleanup. All residue would be landfilled or delivered elsewhere for further processing.

4.2.2 High-Technology Material Recovery Systems

A high-technology material recovery system would be much more capital intensive than the preceding low-technology approach. Such a system would mainly rely on mechanical equipment to process and recover materials. Considering applications of only those processes which are deemed commercially available, the following types of material recovery processes would be possible:

- Magnetic ferrous recovery;
- Film plastic recovery; and
- Paper recovery.

High-technology material recovery systems use mechanical equipment to remove specific materials. Generally, equipment is available to perform the following specific functions:

- Receiving and storage;
- Size separation;
- Size reduction;
- Classification of materials; and
- Materials recovery.

4.2.3 Mixed Waste Processing Systems and Vendors

The following is a list of potential processing equipment or system suppliers who may be willing to furnish all of the equipment for mixed waste processing system.

- AENCO
- Buhler-Miag
- Flakt
- Heil
- Lundell
- PLM Sellbergs
- Sorain-Cecchini
- Valorga

AENCO has not supplied equipment for any projects in the U.S., but they have operated the Albany, NY processing facility since 1981. AENCO's processing system centers around a patented rotary drum air classifier. Recyclable materials are recovered from the heavy fraction of the material separated out by the rotary drum air classifier. The light fraction is screened and shredded to produce RDF.

Buhler-Miag has supplied equipment for several European RDF facilities and for one facility in the U.S. located in Eden Prairie, MN. A Buhler-Miag system typically involves the following processes; screening, handpicking, primary shredding, magnetic separation, air classification and secondary shredding. The undersized materials from the screening process and the heavy materials from the air classification process can be composted, and the RDF can be pelletized for storage and shipping.

Flakt has developed and supplied the equipment for the Hogdalen facility, located in the Stockholm region of Sweden. At the Hogdalen facility, several stages of shredding, screening, magnetic separation and air classification are used along with jigging and froth flotation to separate the waste stream into the following fractions: ferrous metals, paper, plastics, mixed color glass, aluminum, fines and a combination of stones and ceramics. Flakt is reportedly not marketing entire systems at this time, but will supply some equipment associated with the plant.

Heil has supplied all of the processing equipment associated with the RDF and material recovery facility in Portsmouth, VA and the Haverhill/Lawrence, MA facility. Construction of the Portsmouth facility (owned and operated by the Southeastern Public Service Authority of Virginia) was completed in 1987 and the facility has handled approximately 2000 tons of mixed waste daily since commissioning. The processing system consists of primary and secondary trommel screening, secondary shredding, aluminum hand picking, and multi-stage magnetic separation.

Lundell has supplied equipment for small scale projects in the U.S. Lundell systems feed waste to a low speed shredder or bag breaker and then to a magnetic separator. The magnetic separator is followed by a patented organic separator section (essentially disc screen which removes the noncombustible organic waste, glass, dirt and fines), a sorting table for handpicking of nonferrous and plastic materials, and a high speed shredder/densifier to produce RDF. The organic waste residue is sent to a shredder and then composted.

Sorain-Cecchini, an Italian company, has supplied equipment for several plants inside and outside of Italy. Sorain-Cecchini utilizes a proprietary pick-up device and bagbreaking process as well as air classification to separate film plastic from the waste stream. In addition, their system typically includes the following processes to recover ferrous metal, glass and aluminum and to produce a densified RDF: primary and secondary screening, magnetic separation, eddy current separation, handpicking and densification.

Valgora's equipment has been used in a small industrial facility in France. This facility has been in operation since 1984. The Valgora process crushes, screens and classifies the waste to remove ferrous metals and heavy inert materials. The resulting product is fed to an anaerobic digester. The digester material is crumbled, screened and densified for subsequent combustion with a waste heat boiler.

4.2.4 Marketability of Processed Materials

The marketability of material recovered from solid waste depends, in part, on the standards of acceptability of the recycled material markets. When materials are recovered from mixed solid waste, the quality and quantity of the recovered material is reduced. Compactor trucks can crush a large portion of the glass. The crushed glass can contaminate much of the paper with shards, moisture, food waste and other debris. Metal cans and plastic bottles may be crushed, potentially trapping food waste inside resulting in material contamination.

Material upgrading by washing to remove contaminants may be needed before they are acceptable for the recycling markets. The price and marketability of recycled materials (i.e. paper, glass, aluminum and ferrous metal) are determined on the basis of supply and demand.

The introduction of mandatory recycling has resulted in these materials being recovered from the waste stream in far greater volume than in the past. As more and more material enters the markets, the potential exists (and in many cases has occurred) for the saturation of specific markets. The result has been negative pricing and more stringent demand for product quality. Careful consideration of this issue is urged prior to developing mixed waste recovery capacity.

A summary of the material recoverable through mixed waste processing systems is provided in Table 4-1. Brief descriptions of the marketability of the materials recoverable through these processes are provided below.

4.2.5 Recoverable Materials - Paper/Corrugated

Paper is, by weight and volume, the largest portion of most municipal waste streams. The paper recycling market employs strict specifications for material quality. The marketability of various grades depends on current demand and whether it was consumed domestically or exported. The export markets generally have more strict baling and contaminant requirements due to the added costs of shipping. Further, exporters are unwilling to risk having a load rejected on the docks of a foreign country due to contaminants.

Two common specifications for paper recycling relate to moisture content and contamination by foreign material. Moisture is important because paper is bought by the pound. Contaminants in recycled paper include glass, metal, plastic, dirt, foodwaste, clay coatings and adhesives for binding. Contaminants reduce the quality and the quantity of the reprocessed product.

4.2.5.1 Aluminum

Aluminum recycling is the economic mainstay of most recycling systems. The market value of recycled or secondary aluminum, and, in particular, the used beverage containers, depends on several factors. The prices paid are partly dictated by the cost of mining, importing and processing the raw material used in primary aluminum (i.e. the avoided cost). Another factor affecting the value is the level of contamination. Contamination impacts four areas of the recycling and manufacturing process: weight, quality, time and safety.

TABLE 4-1

COMPARISON OF PROCESSED SOURCE SEPARATED MATERIALS WITH MIXED WASTE PROCESSING (MWP) AND RECOVERED MATERIAL MARKETABILITY

Material	Processed Material	MWP Material
Corrugated Cardboard	Yes	Yes
Glass Food Containers	Yes	No
Metal Food Containers	Yes	Yes
Newspapers	Yes	No
Office Paper	Yes	No
Scrap Paper	Yes	Yes
Storage Batteries	No	No
Waste Oil	No	No
Yard Waste	No	No
Plastics	Yes	Yes

NOTE: A mixed waste processing facility could allow for the front-end separation of corrugated on the tipping floor and from the feed conveyor prior to the material entering the processing system. However, paper would not be routinely separated during the process due to contamination and moisture.

As with most recyclables, aluminum is purchased by weight. Contaminants add to the weight and decrease both the quality of aluminum recovered for remanufacture, thereby reducing the marketability of the recycled material. Quality is affected if contaminants intrude on the remelting process. Foreign materials can render a batch of remelted aluminum unusable.

Valuable time can also be lost if contaminants enter the processing equipment. When moisture, along with dirt and syrups from cans combine, they can form a thick, pasty material that can clog the operation and delay the remanufacturing process. Finally, safety is an issue that may be affected by contamination. Water and molten aluminum when mixed can react explosively.

By keeping aluminum cans out of the mixed waste stream, the possibility of picking up contaminants is reduced. Reynolds Aluminum Recycling Company has experimented with MSW separation to obtain aluminum by installing a very simple mixed waste processing type system in Houston, TX. The material recovered in this process has been of considerably lower quality. At the Southeastern Public Service Authority of Virginia facility in Portsmouth, aluminum cans in the mixed waste stream are mechanically sorted then handpicked. The recovered aluminum is sold to Reynolds Aluminum. There are some trapped plastic sheets and paper, but the product quality is acceptable to the buyer. Shredding and washing processes could be added at a mixed waste processing facility to improve quality and quantity of recovered aluminum.

4.2.5.2 Plastics

Plastics are relative newcomers to the recycling materials industry. Expanded use of plastics in packaging and other industries in recent years has caused an increase in the percentage of plastics in the waste stream. The characteristics which contribute to increased use of plastics are now hampering disposal and recycling efforts. In particular, its light weight and elasticity result in large volume requirements for disposal.

The many different plastic resins and the difficulty in their quantification has further hampered recycling efforts. Recovery methods thus far have been limited to specific identifiable plastic products such as high density polyethylene (HDPE), in particular gallon and half gallon liter carry bottles, and polyethylene teraphthalate (PET)(one and two liter average containers). The large scale success of plastics recycling may depend on proper labeling of the various types of resins to render them readily and accurately identifiable. However, source separation in conjunction with the use of trained handpickers at material recovery facilities can accomplish the recovery of plastic containers. Film plastic, such as those used in grocery and garbage bags, can also be recovered at the MRF via handpicking and mechanical means.

Once HDPE and PET have been separated at the MRF, the material can be shredded to allow drainage of moisture and contaminants and to reduce volume. Washing after shredding can decrease contamination. Shredding is especially important if caps are allowed to remain on the containers as compaction is difficult with capped containers.

4.2.5.3 Glass

Glass manufacturers have been using recycled glass for years. By recycling glass they are able to reduce the consumption of energy by reducing furnace temperature needed to melt the material used in making glass.

The extent to which glass must be prepared for glass remanufacture depends on the end use and downstream processing equipment. If glass is to be recycled for use in the glass packaging industry, it must be color separated and free of contaminants. Further processing requirements are determined by the company buying the glass. Some companies have the capability to remove metal rings, others do not. Some may accept only whole containers and others may accept only furnace-ready, color separated cullet.

Glass mixed with municipal solid waste and loaded in a compactor truck (using full compaction capability) will likely break before recovery. In fact, many source separation based, material recovery facilities report glass breakage rates of greater then 50 percent due to excess handling, transfer, and conveyor turns in material recovery systems. Once the glass is broken and mixed with other colors and small debris, it becomes difficult to separate and market. If the glass fractions could be separated from other dirt and debris, there may be a market for it in the construction material industry as fiberglass or road aggregate. This market is limited and the cullet price is generally half that paid by the glass packaging industry. Therefore, a mixed waste processing system would not normally be designed to recover color separated glass.

4.2.5.4 Ferrous Cans

Ferrous cans are comprised of three general types. The most abundant type of container is the tin-plated steel food container. This container is a steel can that is coated with a thin film of tin. A second type of containers is a steel can that does not contain the tin coat. A third type of container is known as the bi-metal can. This container, used primarily for beverages, has a bottom and sides made of steel with an aluminum top.

The tin plating must be removed from the tin-plated steel can before it can be recovered for use. In some cases. the steel industry accepts small quantities of either tin-plated or bi-metal cans directly from scrap dealers or municipal recycling programs. These containers should be baled or densified (crushed). Detinners (companies which remove and recover the tin) often require that the tin cans be free of labels and thoroughly rinsed of contaminants. The cans are often shredded in one or two-inch squares to reduce volume. Alternately, the containers are flattened and shipped loose with labels attached. Again, the price paid is partly determined by the level of contaminants.

4.3 Mixed Waste Composting

Mixed waste composting has much in common with mixed waste processing discussed above. A mechanically oriented "front end" process precedes most mixed waste composting facilities. These facilities are designed to convert the organic fraction of a mixed municipal waste stream to a marketable soil conditioner product. Mixed waste composting is a natural process whereby the organic portion of the waste stream biologically decomposes under controlled conditions. Table 4-2 lists the compostable components of the "typical" waste stream as well as their corresponding percentages of the total waste stream. The final product of the composting process is a humus-like material, dark in color with a crumbly texture and earthy odor.

4.3.1 Composting Processes

Composting can be carried out under either aerobic (oxygen rich) or anaerobic (lack of oxygen) conditions. Generally, anaerobic composting is contained in a sealed vessel and involves mechanical agitation, and takes place at relatively low temperatures. Anaerobic composting takes much longer than aerobic composting and methane gas is formed as a product of the biodegradation process. In some third world countries manure is anaerobically digested with the methane used for electrical power production.

Aerobic composting requires the presence of moisture and oxygen and produces carbon dioxide as a byproduct. This process is characterized by relatively high temperatures (150 degrees F) and more rapid decomposition than an anaerobic process. The higher temperatures allow for the greater system capacity. For these reasons, aerobic composting is the common method used in the U.S. for MSW composting and is the type that will be discussed in further detail in this section.

TABLE 4-2

MATERIALS IN THE MUNICIPAL SOLID WASTE STREAM

Compostable Material	Percentage of Total
Paper and Paperboard	40
Yard Waste	18
Food Waste	7
Wood	4
Miscellaneous Organics	1
Subtotal	70

Noncompostable Material	Percentage of Total
Glass	7
Metals	9
Plastics	8
Rubber and Leather	3
Textiles	2
Miscellaneous	1
Subtotal	30
TOTAL	100

Note: Figures are based on typical solid waste stream without a major source separation program in place. Source separation will reduce paper and paperboard, glass, plastic, and metal components depending on the level of recycling.

Source: Franklin Associates, Ltd., 1992.

Most composting systems consist of four basic steps:

(1) preparation of the feedstock by preprocessing;

(2) digestion of the feedstock (composting);

(3) curing; and

(4) postprocessing.

The preparation steps can include sorting solid waste into organic and inorganic fractions via mechanical and/or manual materials processing, reducing the size of the feedstock, adjusting the moisture content of the feedstock and adding nutrients and bulking agents. The digestion step is the critical step in which much of the biological action takes place, and the curing step allows for the completion of biological activity. Finally, postprocessing (which may involve screening of the compost) completes the process and prepares the product for market. The duration for the entire MSW composting process to be completed varies from one type of system to another and can range from 6 to 15 weeks depending on the intended markets for the compost. The finished compost product represents a 25 to 65 percent reduction in volume from the incoming waste.

4.3.1.1 Feedstocks

Almost any biodegradable organic material can be a suitable feedstock for composting, although materials whose carbon is available in simple organic molecules to the microorganisms result in a higher grade compost product and are more quickly processed.

When solid waste is converted into compost, extensive feedstock preparation is required. The inorganic and non-biodegradable materials in the solid waste products need to be removed to improve marketability of the compost product.

Composting of the biodegradable fraction of solid waste normally results in approximately one-third reduction by weight due to evaporation and decomposition. Rejects and screenings from processing the compost material are also approximately 25 to 35 percent of the feedstock stream. Reject quantities are dependent on the level of source separation and preprocessing to remove inorganics.

4.3.1.2 Preprocessing

Preprocessing of solid waste typically includes receiving, previewing and sorting, material screening, magnetic separation, grinding and the addition of moisture (i.e., water, sewage sludge, canning wastes, etc.). Sorting involves the removal of large noncompostable or bulky items, household hazardous wastes and salvageable materials such as tires, large pieces of wood, metals and certain papers. By removing glass, metals and other abrasives, the sorting process protects the composting machinery, improves the quality of the final product, and facilitates salvage and recycling. Most of the sorting can be accomplished by manual separation or screening techniques, while ferrous metals are removed in the preprocessing or postprocessing stages by magnetic separators.

Grinding or shredding reduces the particle size to facilitate handling and digestion. Not all processes use grinding prior to digestion. In the Dano process, waste is ground by the glass, metals and other inorganic materials as the waste is tumbled in the digester. Further separation such as screening may also be done before or after grinding to remove large pieces of materials and noncompostables or to recover combustible materials for use as a fuel.

Prior to digestion, the moisture content must be adjusted to ensure proper process conditions. Optimum moisture for digestion is generally in the range of 50 to 60 percent. The moisture content of unprocessed solid waste typically ranges from 15 to 25 percent. To adjust the moisture content of the refuse, water, raw or digested sewage sludge, animal manure, canning wastes or other sources of moisture may be added. Sewage sludge, manures and other waste may have added advantages, including additional organic and trace materials as well as microorganisms, all of which contribute to the decomposition process and to potential for aiding in plant growth. Sewage sludge, however, may have the disadvantage of containing heavy metals and other toxins. Any process which uses sewage sludge is also subject to more stringent regulations. Regulations regarding MSW compost are in the early stages of development and may change as actual operating experience is gained.

4.3.1.3 Composting

The process of composting can be represented as a chemical reaction:

$$\text{Fuel} + \text{Microbes} + O_2 + H_2O \dashrightarrow H_2O + CO_2 + \text{Ash} + \text{Energy}$$

The fuel must be organic in order for it to be broken down by microorganisms. The microbiological population is usually present in ample amounts in raw garbage and even more so in sewage sludge. Oxygen and water are supplied by either turning (similar to tilling) or air injection depending on the system used. The by-products are water vapor, carbon dioxide, residue (the inorganic fraction inherent in the feed materials), and energy in the form of heat.

The fuel quality and process control are the key to an efficient process. Energy from the process is a result of the digestion of carbon by the microbes. Therefore, a good fuel must have ample supplies of carbon and the carbon must be easily accessible to the microbes. For this reason, the simple molecular structures found in food waste makes it an excellent fuel. Paper and wood products, which are high in complex cellulose and lignin, are not as easily accessible to the microorganisms and therefore take a longer time to decompose.

The process is controlled by maintaining the appropriate temperature and moisture content to maximize the organic decomposition. Microbial action is inhibited by temperatures greater than 140 degrees F to 150 degrees F, but temperatures must be maintained above approximately 125 degrees F to 130 degrees F to destroy pathogens present in sewage sludge or raw waste. Moisture content becomes a limiting factor when it drops to below 60 percent to 50 percent, and microbial activity becomes severely inhibited when the moisture drops to 25 percent. Operating experience has shown that moisture should be maintained at approximately 55 percent during active composting.

Another important element of the process is nitrogen, a major constituent of microbes without which, the multiplication and proliferation of microbes will not take place. The ratio of carbon to nitrogen (C/N) is critical to the rate of decomposition. The optimum C/N ratio is within the range of 19 to 30 parts of available carbon to one part of available nitrogen. Table 4-3 shows some materials which can be used to adjust the C/N ratio to optimize the composting rate.

4.3.1.4 Curing

The curing phase provides additional time for the compost digested in the initial stages of the composting process to stabilize prior to packaging or delivery to a market. The material is piled in windrows which are turned on a less frequent basis (every 6 to 8 days) compared to the digestion process. The amount of curing necessary depends on the specific market for the compost. If fresh or insufficiently digested compost with high C/N ratios is applied to soil, the continuing microbial action could, in theory, deplete the soil of nitrogen. For this reason, compost from a high-rate digestion process typically undergoes a curing period which allows for further digestion.

TABLE 4-3

NITROGEN CONTENT OF MATERIALS

Material	C/N Ratio
Low C/N Ratio	
Horse Manure	20:1
Cow Manure	20:1
Grass Clippings	29:1
Food Waste	15:1
Humus	10:1
Sewage Sludge	
Digested	6:1
Undigested	16:1
High C/N Ratio	
Foliage	40:1 up to 80:1
Paper	170:1
Sawdust	500:1
Wood	700:1
Tree Bark	100:1 up to 130:1

Note: C/N = Carbon/Nitrogen

4.3.1.5 Postprocessing

Postprocessing includes finishing, storage and preparation for the market after the composting and curing processes. Finishing can include screening, grinding or other processes to remove such materials as plastics, glass, textiles, wood and metals which remain in the compost and which might be objectionable to the market. Finishing can also be done prior to curing. The degree of postprocessing performed will depend on the market requirements. Additional upgrading, such as pelletizing or fortifying with fertilizers, may be required to satisfy a particular market. If the compost will be sold retail, a bagging operation may be utilized, otherwise it can be loaded directly into trucks and sold in bulk form. If the compost is to be used as landfill cover, little or no postprocessing is required.

4.3.1.6 Residual Handling

Residuals from composting facilities include materials removed during the preparation of waste as a feedstock and any materials which are rejected during the postprocessing of the compost. The quantity of residuals is dependent on the percentage of the organic component; the inerts such as glass or plastics, in the waste stream; the degree of source separation prior to waste delivery and materials recovery at the composting facility. For most facilities which recover recyclables as a preprocessing step, the residuals which require landfilling usually amount to approximately 25 percent by weight of the incoming waste.

4.3.2 Technologies and Vendors

A large number of manufacturing firms provide equipment for yard waste collection and chipping as well as for windrow turning and air injection for static compost piles. Several of the more widely publicized European and American composting systems now being proposed for use in the U.S. are summarized below. A number of other Japanese and European systems exist but have not established offices in the U.S., and do not actively market solid waste compost applications.

Buhler-Miag preprocesses to recover a fuel and a ferrous metal fraction, followed by shredding mixing in a horizontal homogenizing mixing drum and then windrowing for digestion; bottom aeration and periodical turning of windrows.

Daneco preprocesses to remove recyclables and non-processibles, followed by shredding for size reduction; trommel screen separation, overs are rejects, unders are screened a second time. Wet separation maximizes removal of the organics from the unders stream. Composting takes place in aerated static piles for 8 weeks in a totally enclosed area to control moisture. After the first 2-3 weeks, the piles are turned. After weeks of composting, the material is separated by size.

Dano uses a rotating drum, slightly inclined from the horizontal; one to five days for digestion, followed by windrowing; no grinding; forced aeration into drum; and underground aeration system for windrow slabs.

Bedminster uses a rotating drum, slightly inclined; one to five days for digestion, followed by windrow curing and storage; no grinding; forced aeration in drum; trommel screening of the compost before marketing.

Recomp processes feedstock by trommel with seven inch holes. Oversized materials are rejected and undersized materials are fed into an inclined rotating drum divided into three sections. Materials spend one day in each section for a total of three days then is transported outside to be formed into windrows. Horticultural product is windrowed for 120 days, agricultural for 40 days. Post processing consists of screening and for horticultural product, destoning.

4.3.3 Operational Experience

Solid waste composting has a long history outside of the U.S. A wide variety of technologies have successfully demonstrated that the organic fraction of solid waste can be biologically reduced (decomposed) to a humus-like material. In Europe, composting is one component in the solid waste management system which also includes recycling, incineration and landfilling. MSW composting has received wide acceptance in countries such as France, Italy, the Netherlands and Austria. Many of these countries have been conducting extensive research programs on compost uses and soil amendment advantages for 10 to 20 years or more. The U.S. solid waste composting industry has, however, been hampered by the large number of unsuccessful plants built in the 1950s and 1960s and by a limited number of additional failures in the 1970s, 80s and 90s.

The driving forces to implement MSW composting in the U.S. include:

- The desire to find alternatives to incineration for waste volume reduction, particularly for communities that generate less than 250 tons per day of waste;

- The difficulty in siting waste disposal facilities that are politically, socially and environmentally acceptable;

- The search for cost-effective low-technology disposal options; and

- A search for less expensive and more easily degradable bulking agents than wood chips and sawdust for use in sewage sludge composting.

The operational experience of MSW composting in the U.S. is extremely limited. There are approximately a dozen operating MSW composting facilities in the country. Most are permitted and designed for less than 100 tons per day. Attempts to process larger quantities of material have been plagued with problems, particularly with odors and in marketing compost products.

4.3.4 General Markets for MSW Compost

The developing and unstable markets for compost and the ability of solid waste compost plants to meet market specification have caused considerable uncertainty with respect to product revenues. The fluctuations are often controlled by the principles of supply and demand. Several of the plants that failed in the U.S. did so because the revenue projections were not realized. Compost markets are expected to be found, among other areas, in commercial landscaping, luxury horticulture, the fertilizer industry, agriculture and land reclamation. Compost can also be used as landfill cover, however, this capability will vary from state to state depending on regulations and experience. In New Jersey, the lack of operational experience has resulted in no specific position as yet. Compost, utilized as landfill cover in New Jersey, will most likely be judged based on the existing engineering related specifications.

The luxury horticulture industry (i.e., potted plants) typically requires a high quality, pure,stable and consistent product similar to peat moss. This market is often viewed as best satisfied by peat moss, compost from yard waste or leaf composts.

Typically, the agricultural industry does not demand as a high a quality compost as the horticulture industry because the material is frequently used for erosion control and top-soil stabilization, or for crops not directly consumed by humans. Agricultural applications often dictate limits on heavy metal concentrations and other contaminants for the protection of human health due to potential exposure via plant uptake. Additionally, limits on glass, plastics, and other non-biodegradable components may also be specified. Land reclamation, which is an emerging market, should have the least demanding market specifications. Because most land reclamation is in remote locations where the existing surface soil lacks most or all of the essential nutrients, microbiology, and structural components to promote growth or prevent erosion, the compost users may benefit from the lack of need to control certain impurities such as metal concentrations not tolerated by other markets.

The only waste compost currently achieving marketing success in the U.S. is sludge compost. Even in Europe, where composting is more extensive, sludge is the major source of waste-generated compost. Municipal solid waste compost in Europe is used primarily for mine reclamation, landfill cover, or municipal green areas such as parks.

In order to market compost, the quality must meet the specific demands of the market sector targeted as well as the specific user. For agricultural applications, acceptable quality may be crop-dependent. Many businesses are reluctant to switch from a proven organic soil additive to something unknown and which may potentially diminish crop quality. Research on the effects of compost is ongoing. Current research has indicated that immature compost has been the cause of failed crop tests. Formation of certain acidic solutions has been cited as one specific cause related to maturity; however, definitions and methods for determining compost maturity vary among all researchers. Another commonly cited problems is the high C/N ratio which leads to the robbing of nitrogen availability to plants. The price that various markets will pay for compost is directly related to the quality of the product, the availability of alternatives such as topsoil, peat, or humus and the quantity purchase arrangement.

Public works projects may be a good outlet for composted MSW. Projects may include roadside embankment stabilization, parks and public facilities landscaping, etc. Communication with various departments of transportation has revealed a willingness to consider the acceptability of compost use. At least one, the New Jersey Department of Transportation, has expressed concern regarding the use of compost because they have had experiences with PCB contamination linked to compost accepted from a Delaware facility.

4.3.5 Compost Regulations

Since composting of MSW has not generally been attempted as a feasible means of landfill diversion until recently, very little consensus has been reached on how it should be regulated. A small number of states have promulgated detailed regulations specifically for the composting of MSW or the usage of compost product. Florida, New York and Minnesota are three examples of states that have moved forward. All three of these states have compost quality standards for metals as they relate to usage of the compost. Table 4-4 summarizes the standards for these examples and for Europe.

4.3.6 Site Requirements

Selection of the site location for the MSW facility will directly influence the potential for environmental impacts. In the case of MSW composting, a considerable area will be required for a significant waste stream. Depending on the composting technology employed, the area requirements could range from between 10 and 40 acres based upon a 100-500 ton per day waste stream. One key factor is the requirement for curing time (3 to 12 weeks). The area required also depends on the waste generation rate, processing requirements, active composting time, the curing time, stormwater runoff handling/processing requirements, leachate treatment and property setback requirements. Multiple sites may be required. Each additional site will required a separate scalehouse, stormwater runoff control, leachate treatment and property setback. Each site will also require its own permitting effort.

Environmental impacts of concern relate to traffic, noise, odor (a primary reason for problems at a number of sites), stormwater runoff and leachate, landfill requirements for residue, and socioeconomic impacts. While considered by many to be a major contributor to recycling goals, the reality of the siting process and environmental evaluation will quickly reveal public opposition of a magnitude associated with a landfill siting project.

4.4 Yard Waste Composting

Leaf and yard waste composting has long been practiced by homeowners who need to dispose of grass clippings, leaves and garden waste. The national average contribution of yard waste to the total waste stream is from 15 percent to 20 percent. Recognition of this fact has led to the implementation of yard waste composting facilities throughout the country. Programs range from individual municipalities collecting leaves every autumn, to county-wide programs which accept all leaves and some clippings.

Processing at yard waste facilities ranges from simply forming the collected leaves into windrows and turning annually, to varying combinations of bag removal, shredding, windrows with forced aeration and temperature monitoring and control. The level of technology selected depends on the quantity of materials to be composted, the area available, and the financial resources which can be allotted to this waste management option. Many smaller municipalities use public works equipment to collect leaf waste and form windrows. The public works department may turn the windrow once per year due to lack of equipment, or it may work with the county, or other municipalities to share resources. The compost is usually given away on a first come first served basis to residents, with the remaining material used by the public works department, other governmental agencies or marketed to commercial consumers.

TABLE 4-4

SUMMARY OF REGULATORY LIMITS FOR HEAVY METALS IN CLASSES OF COMPOST (mg/kg, dry weight basis)

	New York[2]		MN[3]		Florida[4]		Europe[5]
Metal	I	II	I	II	II	III	I
Cadmium (Cd)	10	25	10	15	30	100	3
Chromium (CR)	1,000	1,000	1,000				150
Copper (Cu)	1,000	1,000	500	450	900	3,000	150
Lead (Pb)	250	1,000	500	500	1,000	1,500	150
Mercury (Hg)	10	10	5				3
Nickel (Ni)	200	200	100	50	100	500	50
Zinc (Zn)	2,500	2,500	1,000	900	1,800	10,000	500

Notes: (1) The regulatory limits shown are the maximum concentration allowed for the given class of compost. Usage is restricted based on class designation.

(2) For New York Class I compost can be used on food chain crops and for other agricultural and horticultural uses except that it cannot be used on crops grown for direct human consumption. Class II compost is restricted to use on non-food chain crops.

(3) For Minnesota, Class I compost may be distributed for unrestricted use. The Commissioner of the MPCA determines the restrictions on all other compost.

(4) For Florida, Class I compost has no restrictions. Class II is restricted to commercial, agricultural, institutional or governmental operators. Class III has the same restrictions as Class II and cannot be used where human contact is likely.

(5) European standard for Class I compost (unrestricted use).

To achieve the national average for yard waste composting, grass clippings must be included. When composting grass clippings are incorporated, another list of concerns must be addressed. Grass compost requires more frequent turning and tends to produce odors. Considering where the grass clippings originate, the compost must be sampled for contaminants from lawn fertilizer, pesticides and weed killers. Any traces of lawn chemicals could hamper marketing and limit the uses for the compost.

The yard waste composting process must be controlled properly in order to maintain the optimum rate of decomposition. Process control may be accomplished with a feedback control circuit between temperature monitors and blowers, and prescribed moisture addition regimen. Since leaves are high in carbon, some municipalities add nitrogen containing material to achieve a more favorable C/N ratio which allows a more rapid composting rate.

Of the states discussed in the previous section, only Florida requires sampling of compost made from yard waste. The analyses required are for moisture, nitrogen, phosphorous, potassium, organic matter and pH. Florida assumes that yard waste compost will conform to Metals Concentrations Code 1 and, therefore, does not restrict its use. Similarly, New York and Minnesota do not restrict the use of compost from yard waste, if no septage, sewage sludge or solid waste is used in the process.

4.4.1 Yard Waste Processing Options

The two main yard waste recycling techniques are:

- The composting of leaves, grass and chipped brush; and
- The chipping of woody branches to be used as mulch.

A composting program can utilize the low-technology static pile method of composting, the windrow method, the forced aeration method or some variation of the three. Mulching operations usually utilize a tub grinder for chipping large pieces of wood and a shredder/screening operation for chipping smaller brush type items. Portable chipping machines can also be used to produce mulch as the yard wastes are collected. The collection of yard wastes can be completed using a combination of the following methods:

- Vacuum truck pickup of leaves raked to the curb;
- Curbside collection of bundled and bagged yard wastes;
- Citizen drop off of yard wastes at a waste management facility;

- Semi-automated curbside collection of yard wastes utilizing a cart container system; or

- Trucks outfitted with hydraulic lifts and/or portable-chipping-unit truck combinations.

Composting entails a combination of mechanical and biological processing over a 12 to 24-month time period, whereas the chipping of wood wastes for use as mulch only entails simple mechanical processing.

Decomposition of organic matter is a function of:

- The amount of air in the pile;

- Moisture content of the material;

- Temperatures generated by microbial activity;

- Particle size; and

- Organic and nutrient composition.

Grinding of material reduces particle size and increases the surface area of the material, which promotes faster decomposition. The addition of specific organic materials at specific times during the composting period can aid the decomposition process and enhance the quality of the end product. The addition of grass clippings to a 12 month old leaf compost adds nitrogen to the compost, thereby accelerating the decomposition process, but will increase odor problems.

Aerobic decomposition refers to the biological breakdown of material that takes place in the presence of air. This is preferable to anaerobic decomposition (in the absence of air) which is odoriferous. If the material is coarsely ground rather than finely ground, air will be able to penetrate the piles much more easily. Finer material like leaves and grass will need to be turned or fluffed frequently to ensure an adequate oxygen supply. The addition of wood chips to a compost pile or windrow also allows for greater air penetration.

The moisture content of decomposing material should be maintained at 50% level, although there is considerable leeway. Too much moisture may result in water leaking out as leachate. Too little moisture causes microorganisms to greatly reduce their activity. The main disadvantage to this is that decomposition will take longer.

As the microorganisms work, the pile heats up to temperatures of approximately 140 degrees F. Maintaining these temperatures for an extended period of time is necessary to destroy pathogens and weed seeds.

Time is another factor in the composting process that can vary greatly. The amount of time required for decomposition is dependent on maintaining proper air, moisture, and temperature conditions; material type; degree of decomposition desired; frequency of windrow turning; and available space. Complete composting of yardwaste will in general take from 12 to 24 months with an "active" composting program and from four to five years with a static low technology program.

4.4.1.1 Windrow Method

The most prevalent method of composting is through the use of windrows. By this method, the leaves are delivered to the site in bulk and formed into windrows using a self-propelled windrow turning and aeration machine and/or a front end loader. Windrows are piles approximately 6 to 10 feet high and by 12 to 25 feet wide. The length is limited only by the size of the site. The windrows should be turned and reformed using the front end loader at least twice per year. The main advantage of this approach is that it is relatively inexpensive and produces compost in a practical amount of time. The disadvantage is that it requires a large area in which to complete the composting process.

4.4.1.2 Forced Aeration

A variation of the windrow method is the use of static piles that are not turned, but have air forced into the piles through the use of perforated pipes and either a forced-draft fan or an induced-draft fan. Front end loaders are used to form, manipulate, and load out these static windrows. The advantage of this approach is that less space is required due to the use of large windrows and forced aeration speeds along the composting process. The disadvantages include its higher start-up costs due to the capital costs of the forced-air system, the site preparation required for its use, and higher operation costs due to pipe breakage and pump maintenance.

4.4.1.3 Chipping and Mulching

All of the wood wastes collected can be processed in either a tub grinder, a shredder or a portable chipping unit. This material can be sold as mulch for agricultural and landscape applications. Only separate, clean yard waste (free of metals, etc.) should be processed for sale as mulch. A certain percentage of the wood chips will be required by the composting program as will a certain portion of the grass clippings depending upon the composting program design.

4.4.2 Site Requirements

Siting composting facilities may be less difficult or be less controversial than siting landfills, but does have environmental considerations. For example, proximity to residences or commercial retail centers must be considered in the siting process because of the potential odors. Site drainage and runoff are also critical considerations. The site should not be near wetlands, streams, or other surface waters unless provisions are made to capture any runoff contaminated with potential pollutants. Utilizing a suitably paved site minimizes concerns about erosion or the ability to accommodate heavy equipment but may also increase the requirements for runoff treatment systems. Other siting criteria, such as roadway access, haul distance, topography and access to utilities, are similar to those for other waste management facilities which must be remotely located. Proximity to end use markets may also be a valuable siting criterion.

The land area required for composting is directly proportional to the throughput capacity required. A site will require areas for receiving, processing (decomposing, shredding, screening, and curing), storing and exporting. Setback guidelines (minimum buffer requirements) from property lines, residences and businesses, and surface water must be considered. The site may also require a stormwater runoff sedimentation basin, which will depend on the size of the operation and the topography. A rule of thumb regarding area requirements for a passive, non-aerated leaf compost process is 0.08 - 0.13 acres per 1000 cubic yards of material and from 0.10 - 0.29 acres per 1000 cubic yards of material for aerated processes. The configuration is also affected by the setback guidelines and site geometry. As an example, a facility processing 10,000 cubic yards of compost on a square site using a windrow and turn method (non-aerated process) would require about 1.67 acres for processing area, an additional 15 percent or 0.25 acres for storage, and approximately 3.59 acres for buffer, or 5.51 acres. It may be possible to store and load out cured compost material within the buffer area depending on the site layout, thereby reducing land requirements.

4.4.3 Equipment and Costs

The types of equipment required for composting and mulch production will vary depending upon the finished end products desired by the community. Compost turning equipment can cost on the average $145,000 for larger scale self-propelled machines that operate by straddling the windrows while traveling over them. This equipment uses metal teeth to break up, aerate and redeposit the material in piles. Smaller scale versions are available for between $10,000 and $60,000 and can be mounted on front-end loaders.

The front-end loader is the basic piece of equipment for windrow composting. Approximate costs range from $50,000 to $100,000 plus $10,000 for the claw attachment. It can be used to form the windrows, turn them and move the finished product. Either track or wheel front end loaders can be used. Additional attachments, such as a claw (used for loading and moving leaves) are available with most models.

Forced aeration equipment consists of blowers (a one horsepower blower costs approximately $250) and piping. At least one blower will be required per windrow. Six inch diameter plastic pipe costs approximately $9 per linear foot installed.

Tub grinders come in two main types: a self propelled version, which can easily service a number of different sites, and a stationary model designed to service a single site. These units may have to be modified to suit a particular yard waste program in terms of feed mechanisms and portable storage methods.

4.4.4 Material Marketability

Markets for compost can be divided into four major categories which include growers, processors, wholesalers and bulk users. Growers can include golf courses, nurseries, landscape companies, lawn maintenance businesses and sod and sod service businesses. Processors refine the compost to meet specific conditions and market the material. Processors may include topsoil companies, fertilizer companies or sand and gravel companies. Wholesalers typically purchase and sell bagged compost at garden supply or nursery type businesses. Bulk users utilize large quantities of compost to cover acreage for land reclamation, landfill cover or for park or roadside stabilization after construction. Since most road construction is done by state DOT's, discussions with DOT officials could result in a significant market.

4.5 Construction and Demolition Waste Recycling

Construction and demolition waste (C&D) makes up a significant portion of the waste stream in many communities. A State of New Jersey Solid Waste Task Force reported that in 1990, 3.565 million tons of C&D material was generated within the state. That meant that wood waste, asphalt and concrete along with lesser amounts of "other C&D waste" accounted for 29 percent of the total waste stream.

C&D waste is often not considered part of the municipal waste stream, since it is generally disposed of in separate landfills. Because of the nature of construction materials, C&D waste is substantial, particularly in terms of weight. By including C&D waste in community recycling programs, it may be possible to increase the "denominator" on which recycling rates are based. States like New Jersey allow communities to take credit for C&D recycling toward mandated recycling goals.

4.5.1 C&D Recycling Processes

C&D waste processing generally consists of three primary mechanical operations: shredding, grinding and screening. The shredding operation is generally used for stumps and other wood waste, converting the wood into mulch to prepare it for composting or use as a fuel.

Asphalt and concrete are generally processed through grinding machines for size reduction. Screening operations are used for both wood waste as well as hard materials for separation and sizing. Ferrous magnets are also used to recover ferrous materials, such as nails and reinforcement bar embedded in the wood and concrete materials.

Systems available to process C&D waste generally employ one or more of these processes, along with associated equipment, such as conveyors, dust control equipment and rolling stock. Processing facilities can either be enclosed or open. Tipping floors and loadout areas are integral parts of the systems. Some systems have a moderate degree of automation. In addition, some are designed to be "transportable", enabling movement from one construction or demolition operation to another.

4.5.2 Technologies and Vendors

The recycling of C&D waste has been carried on for years, in response to local markets or other conditions. Internationally, the reconstruction of Europe after World War II spawned the development of a number of types of C&D processing equipment, as well as many C&D recycling operations.

Only within the past 10 to 15 years however, has interest grown in C&D recycling primarily in response to increased landfill fees, dwindling landfill space and increased regulation of C&D waste disposal. Many C&D recycling operations are "in house" performed by C&D contractors. In addition, there are a growing number of system vendors, which offer C&D recycling services on a full service (design, build, own and operate) basis. A sampling of these vendors are described below.

4.5.2.1 Lindenmann/TUC

Lindenmann Recycling Equipment of New York City is an established manufacturer of C&D recycling equipment. Recently, Lindenmann teamed with TUC Consult Ltd. of Switzerland to introduce a C&D waste recycling system.

The design of the system, which generally includes a combination of machines that screen, trommel, sort and crush the C&D waste fractions, depends on the local composition of the C&D waste and local markets. The Lindenmann/TUC process has a patent pending in the U.S. Two 1,000 cubic yard per day facilities using this process are operational in Europe, one in Zurich and the other in Basel.

The Basel facility is one of the first C&D plants which processes C&D waste using a trommel, followed by handpicking and further mechanical processing. Wood and metals are recovered while the residual rock fraction is cleaned for disposal in a lower cost landfill. The Basel plant, which occupies a 50,000 square foot site, has been in operation since 1987.

The TUC plant in Zurich has been operating since 1987. This plant is processing C&D waste for recovery of combustible materials which are burned in a nearby waste-to-energy plant. The remaining rock and dirt fraction is landfilled. While similar to the Basel plant, the Zurich facility includes an air classification system to separate the light fraction from the rocks.

4.5.2.2 Bezner Machinery

Bezner is a German company that manufactures equipment and systems for the C&D materials processing industry. Bezner's systems are being marketed in the U.S. by New England CRINC.

Bezner operates a 40 ton per hour C&D facility in Ravensburg, Germany. The system, which was commissioned in 1988, consists of a series of screens, conveyors, magnetic separators and compacting equipment in a 10,000 square foot building. A volume reduction rate of 70 percent is claimed. The plant also processes commercial and bulky wastes.

4.5.2.3 Winzinger Recycling Systems

Winzinger, of Hainseport, New Jersey, owns and operates two concrete recycling plants, a portable concrete facility and a stump shredding facility. The largest concrete facility can process 2,500 tons per day and recovers a product, which sells for $5.30 per ton and is used for road base construction and landscaping services.

4.5.2.4 Angelo's Crushed Concrete

Angelo's Crushed Concrete of Warren, Michigan provides portable concrete processing plants to large C&D projects. In 1987, the company erected a plant to help reconstruct the John C. Lodge Freeway in Detroit. The facility processed 300 tons per hour over two 10 hour shifts for a seven month period. Recovered products included: concrete for new concrete manufacture and road base construction, steel for scrap recycling and asphalt for asphalt recycling.

4.5.2.5 The Sanitary Landfill Company

The Sanitary Landfill Company is based in San Francisco and is owned by Norcal. It owns and operates a transfer station which processes a portion of municipal solid waste from the City of San Francisco. Seventy percent of the waste received at the transfer station is C&D waste, which is reprocessed for recovery of dirt, rock, wood and scrap metal.

4.5.3 Material Marketability

Local markets are needed for materials recovered from C&D recycling operations, due to high transportation costs and traditionally low prices paid for these materials. Because these markets are local, they are also, by definition, variable, especially with respect to demand. For this reason, a local market study is definitely required before a C&D recycling operation is implemented.

Due to the generally low prices obtained for recovered materials, market prices are not the primary consideration when conducting a market study. Rather, the identification of material users who will be interested in using the recovered materials, on a long term basis, even at very low prices, is the major objective. Markets for recovered wood depend on whether it is converted to a mulch or fuel. As a fuel, wood can be sold for approximately $5 per cubic yard.

Recovered concrete and asphalt can be converted to aggregate for use as stabilizers, road bed material, asphalt production and in fill operations. Prices of around $5 per ton could be anticipated. Recovered ferrous materials can generally be sold to scrap dealers, with the price paid dependent on current market conditions.

4.6 Alternative Evaluation

When considering a variety of program approaches it is important to plan for a structured alternative evaluation process. Given socio-political issues related to developing solid waste management and recycling programs, it is helpful to use quantitative approach with a qualitative ranking system (low, medium or high). The following four general parameters can yield an extensive evaluation that is responsive to most concerns.

- Program Impact
- Economic Nature of the Program
- Program Reliability and Flexibility
- Ease of Implementation

For each general area, a number of quantitative parameters may apply. For example under "Program Impact", subconsiderations could include "Net Diversion rate" and "Number of Materials Recycled". Values should be developed early on to establish "scores" that indicate low, medium or high rankings. For "Number of Materials" a low ranking may include alternatives that accommodate 1 to 3 materials, 4 to 6 could be medium and 7 or more high. Low rankings indicate little benefit and high would indicate greatest advantage for the developing entity.

Under "Program Economics", subconsiderations could include capital costs, net cost per ton based upon current market value and net cost per ton with no market value. Estimates made during the planning phase of the program should be considered "ball park estimates" and used only for the purposes of ranking.

"Program Reliability and Flexibility" should be judged on the basis of types of materials that may be recovered, quantity of material recoverable, product quality, market demand for identified products, technical reliability and system flexibility. Another important subconsideration has to do with the value of alternative facility uses should operations cease.

"Ease of Implementation" relates to public acceptance, land requirements, impacts on other recycling or solid waste management system elements, time to implement and regulatory requirements. Time to implement is often underestimated or extended to the point that projects are jeopardized by opposition momentum.

-5-

MARKETING RECYCLABLES

Merely collecting and processing recyclables is not considered recycling - these recovered materials must be reutilized as the source materials for new consumer or commercial products. Thus, viable markets for the sale of recyclable materials, which are retrieved in curbside collection and drop-off and buy-back centers, are essential for the success of a community's recycling program. One of the most commonly reported barriers facing recycling programs is the absence of markets for recycled products. The prices paid by the brokers or end users of these collected and processed materials help subsidize the cost of the overall recycling program. Over the past three decades, the recycling movement has waxed and waned with the shifts in the oftentimes, invisible economic forces of supply and demand underlying these markets.

The purpose of this section is to point out some of the key market variables of which a community decision-maker should be aware and of how many states have recently attempted to strengthen existing markets and develop new markets through a variety of institutional tools.

5.1 Market Forces

Before discussing specific materials and their markets, it is useful to consider some basic issues which affect the consumption (i.e., availability of end-user markets) of recycled materials. On the whole, the U.S. manufacturing sector has been geared to the use of virgin materials. Product manufacturers, in response to society's consumer attitude, have long been more attuned to consumer convenience (especially since the end of World War II) than to product durability and reusability. However, in recent years, as waste disposal costs continue to escalate and the public is becoming more aware of the beneficial impacts of recycling, more manufacturers are addressing recycling and reuse in a positive manner. Still, though, there are significant barriers which apply to the reuse of reclaimed materials. These limitations can be broadly grouped in three categories:

1. Limitations affecting the demand of materials in industry;
2. Limitations affecting the supply of materials to industry; and
3. Limitations imposed by government.

These limitations are interrelated and greatly affect the supply and demand of the recyclable materials market.

5.1.1 Limitations Affecting the Demand of Recyclable Materials in Industry

Figure 5-1A shows the demand curve for recyclable materials utilized by industry to produce final products. This curve demonstrates how much of a particular recyclable material industry is willing to purchase at a given price. The shape of the demand curve (nearly vertical) demonstrates that the demand for recyclable materials is relatively inelastic, (i.e., does not change) with respect to supply. Demand for recyclable materials is dependent largely upon production capacity. Other factors affecting industries demand for recyclable materials are manufacturing capital costs, potential contamination problems, and other costs associated with using reclaimed materials.

Manufacturing Capital Costs

Many manufacturers have complete "in-house" manufacturing processes from virgin materials procurement to final product shipment and may be unwilling to alter these processes in order to accept recyclable materials, for which they have less control over quality, quantity and availability. The expense of current technology has prevented many industries from retrofitting their existing manufacturing process, currently using virgin materials, to processes which utilize reclaimed raw materials. High capital costs required to expand production capacity in order to accept more recyclable materials has also limited the demand.

Potential for Contamination

Raw material specifications can often be met by both virgin and recycled materials; however, the buyer, wary of potential contamination problems, may impose stricter standards before purchasing recycled materials. Large quantities of virgin raw materials that meet material specifications can be procured from a single source, where as reclaimed raw materials must first be collected from a large number of sources, processed and distributed to a few manufacturers. As the number of sources increases, so does the possibility of contamination.

Other Costs

Other costs associated with the use of reclaimed raw materials are increased inspection costs which are accrued during every step of the manufacturing process from collection to final product approval. As a result of the increased production costs associated with reclaimed materials, many of the products produced are more expensive than those produced from virgin materials. In today's economically driven society, there is seldom a large demand for an equal, but more expensive product. However, many recent polls indicate that the public is becoming more aware of the beneficial impacts of recycling and would prefer to buy products packaged in containers made from recycled products.

FIGURE 5-1

SUPPLY AND DEMAND CURVES FOR RECYCLABLES

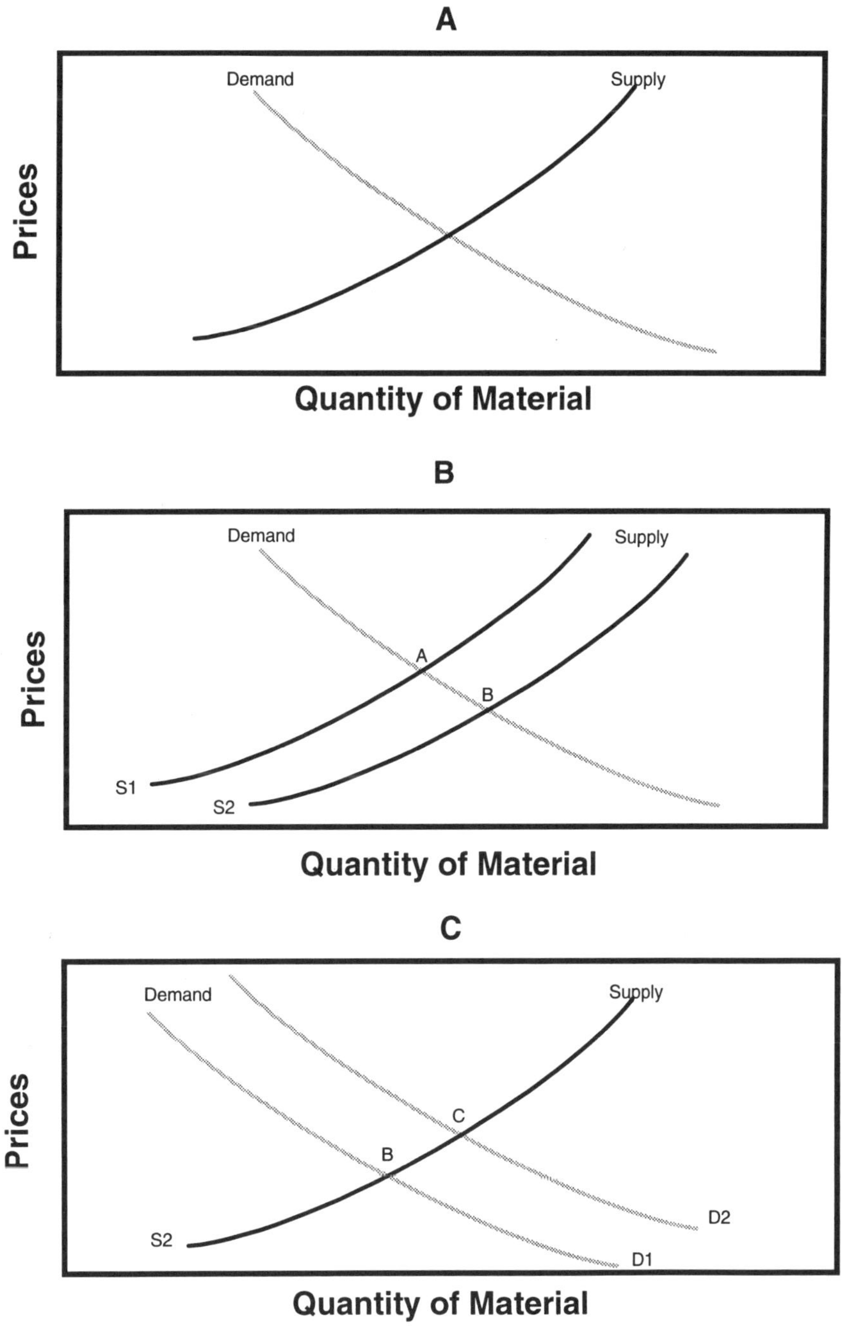

5.1.2 Limitations Affecting the Supply of Recyclable Materials to Industry

Figure 5-1B shows the supply curve for recyclable materials delivered to manufacturers. This curve indicates the quantity of recyclable material suppliers are willing to supply at every possible price during a given period of time. The supply of recycled materials is shown to be extremely elastic because it is relatively easy and inexpensive to get into the supply side of the business. These elastic and inelastic characteristics of supply and demand combine to cause frequent, and often volatile shifts in material prices and supply quantities over short time periods. The recent decline in the price of recovered newsprint in the eastern U.S. is a good example. Also, waste streams vary seasonally. It follows, then, that the availability of certain recycled materials also fluctuates. For example, more beverages are consumed in summer months than in winter months. This results in a higher amount of glass and aluminum beverage containers available for capture and reuse during the summer months. This type of fluctuation may make recyclable materials less attractive to an industry that is looking for a stable, consistent supply of material.

Another limitation facing recycled material suppliers are transportation costs. Natural resources occur in concentrated form, whereas recycled secondary materials from waste are dispersed and have high attendant collection costs.

5.1.3 Limitations Imposed by Government

Up to this point in the discussion, the recovering and reuse of recyclable materials could be thought of as a simple supply and demand interaction, where the supply of recyclable materials delivered to manufacturers equal the demand for those recyclable materials. Point "A" on Figure 5-1B represents a typical equilibrium point. As recently as five years ago, many recyclable markets were felt to be at or near their equilibrium points.

Since that time, however, a great deal of state and federal legislation has been enacted which establishes recycling goals for recyclable materials. To meet these newly mandated recycling goals, communities across the country have established recycling and public awareness programs to promote recycling and reuse. The result of these programs was a dramatic increase in the amount (i.e., supply) of recyclable materials available to an industry which did not (and still does not) have the production capacity (demand) to utilize the materials in many cases. This set of events is shown graphically in Figure 5-1C. Here the original supply curve S_1, is shifted right to S_2, and the equilibrium point moved from point "A" to Point "B". The new equilibrium point B shows a slight increase in the quantity of materials reclaimed and a dramatic drop in price paid for those materials.

To counteract this ominous trend, many state legislatures are beginning to enact legislation to build a recycling infrastructure to help stimulate the market for the purchase of recycled materials. In many states, this legislation has involved development of a multi-faceted strategy which has included grant programs, investment tax credits, loan guarantees, start-up and expansion loans, changes in permit programs for the siting of recycling facilities, information and market clearinghouses, technical assistance programs including workshops and market studies, and changes in state and local procurement requirements which encourage buying recycled products. Another major market impact by government can be the imposition of overall state recycling goals and specific product recycling levels. The intent of all these government is to provide the economic incentive to industry to expand its production capacity by shifting the demand curve for post-consumer recycled products to the right. Thus, increasing material supplies resulting from recycling collection programs will have viable markets as recyclable materials prices will be driven slowly upward.

5.2 Recyclable Materials

5.2.1 Waste Paper

Waste paper, a significant portion of the solid waste stream, is bought and sold on the basis of grade, and prices vary accordingly. Grades of paper range from low grade, such as newspaper and corrugated, to high grade, such as printing, writing, and computer paper. Mixing different grades lowers the quality by reducing the value in remanufacturing. Paper grades are generally defined as specified by the Paper Stock Institute of America which lists specific guidelines that define different grades of paper based on type and preparation. In general, the source of the secondary fiber will dictate the paper grade into which it can be processed.

Paper and paperboard products are distinguished by the physical properties they possess. Generally the physical properties of paper products are dependent upon two primary factors: fiber length and pulping method. Fiber length determines characteristics such as strength, stiffness, opacity, and printability. Basically, as the ratio of long fibers to short fibers increases so does the strength; however, this increase in strength is also accompanied by a reduction in surface smoothness. This becomes a disadvantage from a printing standpoint because ink prints more uniformly on a smooth surface. For this reason, packaging materials (e.g., corrugated) are composed of long fibers while printing and writing papers are composed of short fibers.

The pulping process controls characteristics such as cleanliness and brightness (whiteness). There are three types of pulping processes: mechanical, chemical, and a combination of the two. Pulp produced from mechanical pulping (groundwood pulp) produces a weak paper and usually required the addition of some chemical to hold the paper together as it travels through printing presses. Groundwood pulp contains lignins which provides high bulk and opacity, as well as smoothness and ink absorbency which makes groundwood pulp ideal in the manufacturing of newspapers. Incidentally, it is the lignin that turns yellow when a newspaper is exposed to the sunlight.

Chemical pulping removes the lignin during a cooking process, and as a result, produces a strong pulp that can be bleached to high levels of brightness. Bleached chemical pulp, known as kraft pulp, is used in the production of fine white writing paper. Unbleached chemical pulps are used in the production of brown paper bags and corrugated boxes.

The United States is the largest producer and consumer of paper in the world. There are approximately 700 mills producing paper products in the United States. Two hundred of these domestic mills use only waste paper to make products, and an additional 300 mills use between 15-25% recycled material.

5.2.2 Old Newspapers

Old newspapers (ONP) are a major fiber resource for the paper industry, both domestically and abroad, and are bought and sold as a commodity. On the average, almost every household in the U.S. receives one newspaper daily. The U.S. consumes about 12 million metric tons of newsprint a year with about 60 percent supplied from Canada. Newspaper is the most visible form of waste paper found in the household and is, by both weight and volume, one of the highest percentage (generally 10%) of most residential waste streams. Commercial establishments also generate large volumes of newspaper, but the percentage in commercial waste streams is much lower than in residential waste streams, and is not as easily recovered.

To a household consumer, newspapers consist of newspaper, supplements, magazines, and retail store tabloids. To the end users of recycled newspaper and domestics and foreign paper mills, newspaper has a different meaning. They consider old newspaper (called ONP) a raw material and the many inserts as contaminants. Whether the material is wet or dry and baled or unbaled are also criteria by which they judge ONP. When baled, ONP should contain less than 5% "other paper". Prohibitive materials may not exceed 0.5% and outthrows may not exceed 2%. Prohibitive materials are contaminants, such as glass, plastic, metal, or other non-paper material. "Outthrows" refer to types of paper other than the one being specified. These guidelines show that recycling old newspaper requires a careful effort to produce quality reusable material.

Other paper products such as magazines, telephone books, glossy inserts, and junk mail present several other problems. During the recycling process, ONP is reduced to a pulp by adding warm water and de-inking chemicals. This pulp is then washed and screened to remove any rags, glass, plastics, and dirt. Newspaper normally loses between 15-20% of its weight due to the clay-coated paper on which they are printed. The clay coating separates from the paper during the process and is washed away as waste. The combined loss could be 30-40% of the weight of original fiber. The latex adhesives in magazines and telephone book bindings create latex balls in the re-pulping process which cannot be removed by screening. They can cause spots on the finished product or may cause the paper to stick together in a paper roll.

Old newspaper is recycled into many products. It can be recycled into newsprint and paperboard which is used in production of game boards, book covers, boxes, photo albums, tubes for toilet paper and paper towels, and construction materials such as roofing felt paper, insulation, and wallboard. ONP has also been used for products like cellulose insulation, animal bedding; however, those markets are small regional markets and are not expected to increase in size.

ONP is the main grade of waste paper collected from households. During the 1970's and early 1980's, scout troops, schools, religious, civic charitable organizations and a few private individuals collected old newspapers and sold them to raise funds for local projects and programs. During this period of time, the domestic consumption of recycled ONP fluctuated between 2.4 and 2.9 million tons annually, which corresponded to a recovery rate of 30 percent.

In the mid 1980's, increased demand by a few recycled newsprint mills and an expanding export market caused an increase in the volume and price of recovered ONP. The volume of recovered ONP continued to increase from 4.3 million tons in 1986 to 5.4 million in 1989. This dramatic increase was the direct response of mandatory recycling legislation, and resulted in the collapse of the market prices, as illustrated in Figure 5-2.

In 1991 and 1992 mills within the United States increased capacity to accommodate another 450,000 tons of ONP annually. This increase in capacity has had little affect on the market price of ONP because demand for recycled newsprint in the early 1990's is down. The weak economy has not only reduced the number of newspaper subscribers, but also has decreased the size of most newspapers as a result of reduced advertisement. Although most publishers are looking to purchase recycled newsprint, the fact is, there currently exists more newsprint on the market than the newspapers can consume in many parts of the United States. Recyclers who were formerly buying newsprint have found themselves having to charge a disposal fee on more than one occasion to take the paper in order to make a profit.

To increase the usage of ONP in recycled products, state governments have established mandatory recycled content requirements for many paper products or entering into voluntary agreements with newspaper publishers. Florida was the first state to provide an incentive to use recycled newsprint by waiving a 10 cent per ton disposal tax on recycled newsprint. Legislation in California and Connecticut have established strict standards for the use of recycled newsprint. The Connecticut legislation requires newspapers to increase their use of recycled newsprint from the estimated 20 percent in 1993 to 90 percent in 1998. In New York, a voluntary agreement rather than state legislation requires publishers to increase their use of recycled newsprint from seven percent to 40 percent by 2000. Further, many local, state, and federal agencies, as well as many private firms, are establishing self-imposed purchasing requirements, which require a certain percentage of their paper products to be made of recyclable material. In the past, industries have been fearful to invest millions of dollars into producing products made from recycled fibers since markets for these products did not exist.

Over the next five years, it has been projected by the paper trade associations that ONP capacity will grow significantly as new or retrofitted mills come on line. Known confirmed U.S. recycled newsprint paper mill projects for 1993 will add an additional 400,000 tons of annual recycled ONP capacity. Additionally, several of the major solid waste haulers (e.g., Waste Management Inc. and Browning-Ferris Industries) have entered into sole supply contracts with producers of boxboard and newsprint to sell their collected newspaper. It is currently unknown how these agreements will affect the newsprint market.

5.2.3 Old Corrugated Cardboard

Old corrugated containers, referred to as OCC, represent the largest single category of waste paper collected for recycling. In the U.S., OCC comprises over 40% of all waste paper recycled, and in some large metropolitan areas, over 60%.

To understand OCC, the word "corrugated" means "to form or shape into wrinkles or folds or into alternative ridges or grooves". This definition explains the center portion of OCC or what is known as the fluting or corrugated medium which forms a continuous pattern of grooves. The fluting is placed between two sheets of paperboard to provide strength to a container. This material is commonly called cardboard.

FIGURE 5-2

HISTORICAL PRICING TRENDS OF OLD NEWSPAPER

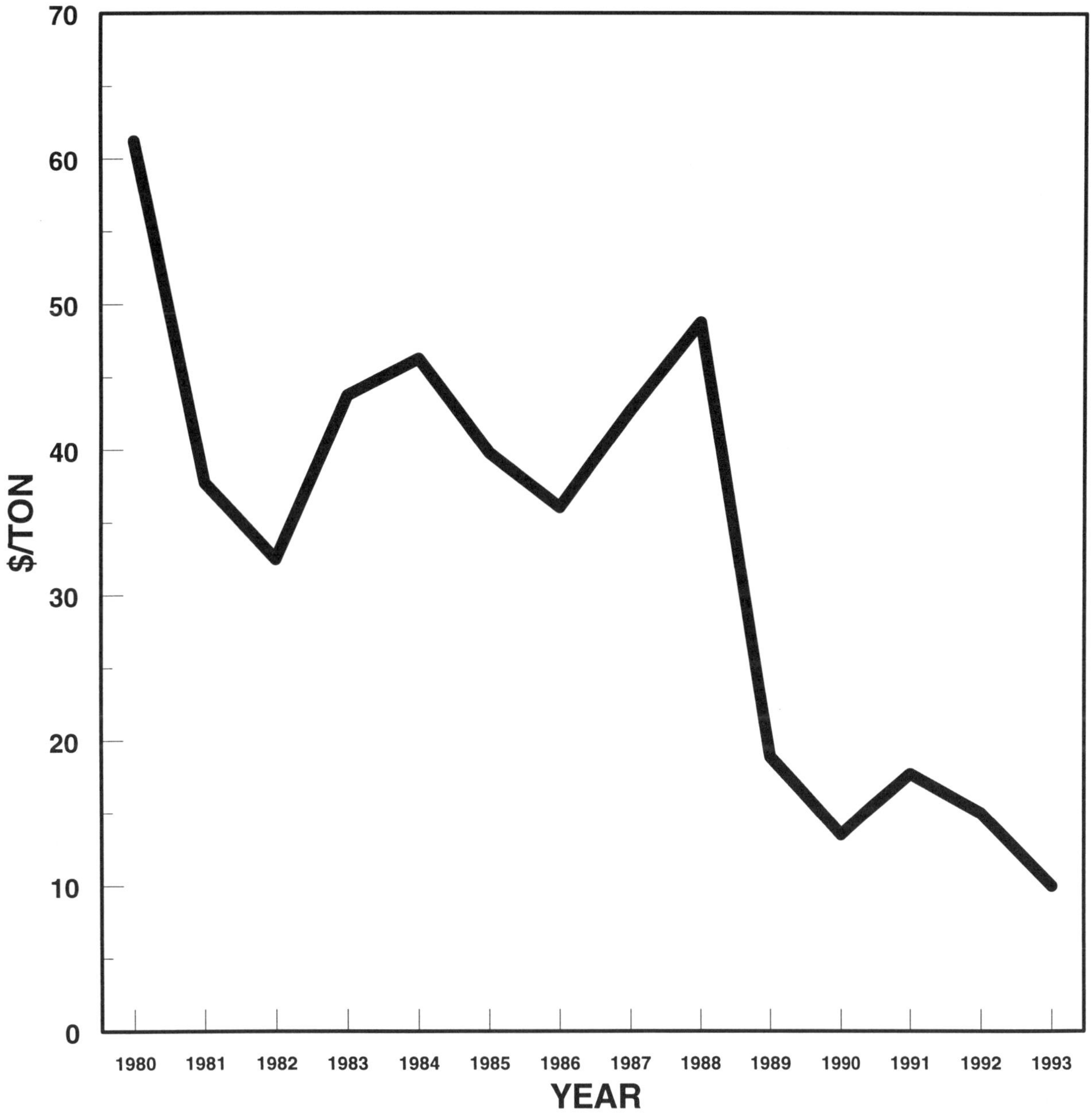

Source: Pulp & Paper Factbook, 1991 and 1992
Recycling Times, 1993
Reflects prices paid on street by processors

Corrugated containers are a major source of waste in a municipal waste stream. Most OCC is generated by commercial establishments such as supermarkets, restaurants, department stores, and various other retail stores. Many large stores keep the OCC separate and sell it to paper dealers or bale it themselves and sell it directly to the mills. The residential segment generates only a small portion of the OCC in the total waste stream, and most of it is contaminated after coming in contact with the other mixed wastes.

Corrugated, like all waste paper used in recycling, must be kept dry and free of contaminants. Both of these requirements must be met if a recycling effort is to realize the maximum value of the OCC. Moisture is a factor that affects the weight of the corrugated cardboard. Prices paid for corrugated are generally by the ton, therefore, OCC must be dry for an accurate weight.

Because most of the corrugated generated for recycling comes from retail establishments, common contaminants are plastics such as trays used to package food items; styrofoam packing materials; plastic bags, wrap, film, and cups; metal objects such as wire hangers, case strappings, can, nails, etc.; plastic and wax-coated cartons such as those commonly used to pack fresh produce (distinguished by a very dark brown color and shiny surface); and other contaminants, including floor sweepings, wood, food waste, beverage cans, trash, etc. Contamination is a serious concern of the recycling industry, since unwanted materials adversely affect production efficiency and finished quality.

Corrugated is used primarily in production of paperboard products for packaging where the strength of boxes, cartons, and fiber cans is very important. Paperboard production is divided into four main categories: (1) unbleached kraft used for outer facing of corrugated and solid fiber boxes, (2) semi-chemical used primarily for center fluting in corrugated boxes, (3) solid bleached paperboard which is converted into packages such as milk cartons, frozen food cartons, and containers for moist, liquid, and oily foods, and (4) recycled paperboard used in folding cartons and set-up or rigid boxes.

Unbleached kraft paperboard has traditionally been a virgin fiber product that contained very little waste paper. However, waste paper use in unbleached kraft mills has increased in recent years from 2% in 1970 to about 9% of the 19.2 million tons produced in 1988. OCC and corrugated box plant clippings are the predominantly used recycled fibers in paperboard production. Continued research into the use of corrugated in unbleached kraft paperboard indicates that this sector of the industry will account for a significant increase in the use of waste paper in paper products. The use of waste paper in this category is projected to increase rapidly to as much as 16% by the year 2000.

Semi-chemical paperboard has traditionally required the use of corrugated waste paper in the range of about 20%, virtually all of which has been from box plant cuttings. The demand for corrugated waste paper in semi-chemical paperboard had increased to 32% of total fiber in 1986. This is a much higher ratio than most experts expected and is due to the installation of cleaning equipment for processing post-consumer waste paper. The use of OCC in semi-chemical paperboard is expected to increase slightly in the future as more mills add cleaning equipment.

Bleached paperboard is solid white throughout and made from primary fiber with little or no waste paper used in its manufacture. Since the largest market for bleached paperboard is in packaging liquid, moist, or oily food, sanitary requirements are of prime importance. Waste paper in bleached paperboard production is likely to come only from clippings and cuttings of paper and paperboard that have been handled, treated, and stored in a clean, sanitary manner. Recycled paperboard traditionally has been the only paper product made from 100% waste paper.

According to the American Paper Institute (API), the recovery rate for OCC was nearly constant at 44 percent for 1980-85, and market prices received for OCC were also constant. However, as the demand for OCC was increased due to increased production capacity, the market prices began to increase from $22 per ton to nearly $60 per ton in 1987. As the price paid for OCC continued to rise so did the number of collection programs. This resulted in a glut of OCC available to the markets, and a dramatic decrease in price (Figure 5-3).

Between 1970 and 1988 the export demand for OCC has averaged roughly 20 percent of the OCC collected in the United States. Japan and Korea are among the largest importers of U.S. waste paper, which they use to make corrugated containers in which to ship TV's, stereos, VCR's, etc. to foreign markets, including America. But due to the economy, demand for these "luxury" products is low. As a result, the export market for OCC and many waste paper products is now extremely low.

Pricing for secondary paper, such as OCC, in the future is hard to predict, due to the many factors that affect the waste paper market (e.g., supply of virgin pulp, demand for paper products and exports, and the effect of increasing governmental involvement in recycling). However, many industry observers believe that OCC market prices will increase in 1994 when the capacity of new mills finally matches the supply of OCC available.

FIGURE 5-3

HISTORICAL PRICING TRENDS FOR CORRUGATED CARDBOARD (OCC)

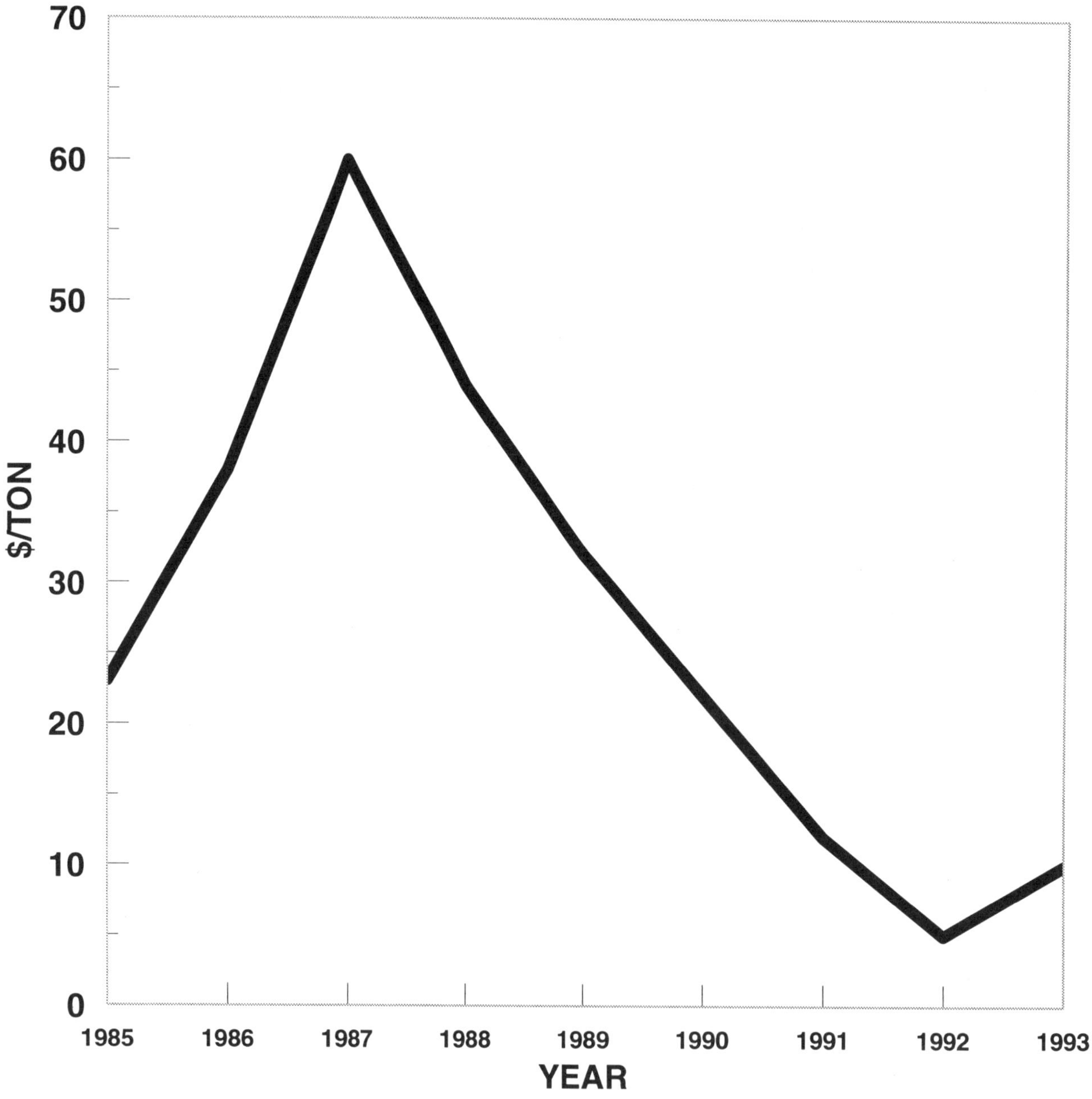

Source: Recycling Times, 1993
Reflects prices paid on the street by processors

Currently, mills are spending millions of dollars to retrofit their facilities in order to accept and use greater percentages of secondary fiber. In 1992, the Seminole Kraft Corporation mill in Jacksonville, Florida spent $110 million to convert its operation to a 100 percent recycled linerboard operation. The mill consumes nearly 550,000 tons of OCC annually.

Thirteen other mills have also been identified throughout the U.S. that will increase OCC demand by another 1.5 million tons annually starting in late 1993. Industry observers have indicated that more mills are seriously planning expanded secondary fiber operations, but have not announced their intentions until contractual arrangements for OCC can be assured. This sudden increase in secondary fiber usage may be the result of an U.S. Environmental Protection Agency recommendation, which calls for corrugated containers to have 40 percent total recycled content, of which post-consumer materials must account for 35 percentage points.

5.2.4 High Grade Paper

High grade paper, also known as office white paper (OWP), is made of long, high quality fibers. This grade can be further divided into two categories: pulp substitutes and de-inking high grades. Pulp substitutes are clippings and shavings from items such as envelopes, bleached paperboard cuttings, business forms, ledgers, and other high quality fibers generally derived from computer paper centers, print ships, and other paper converting plants. These types of high grade paper can be used as a direct pulp substitute for virgin material.

De-inking high grades are usually paper of the same high quality fiber as pulp substitutes, but they have gone through a printing process. They are generally derived from printing plants and office buildings that use high grade paper such as ledger or other forms of high fiber bleached paper. This paper must first go through a de-inking process before it can be used as a pulp substitute. Although high grade waste paper represents a small portion of the total export tonnage, it is in greater demand because of its high quality and, therefore, commands a higher price per ton.

High grade clippings and cuttings as well as de-inking high grades are usually reprocessed back into their original forms. For example, white ledger clippings are re-pulped and used to make new white ledger paper. Another common use is in tissue paper.

The value and recyclability of high grade paper depends on it being separated by grades based on fiber length and on its being free of all unwanted material. The key to recovery and sorting high grade paper lies in producing mill-specific paper bales. Contaminants are dependent upon how the mill plans to reclaim the fibers. Different cleaning and/or de-inking systems are used to produce different paper products. As a result, what is considered a contaminant at one mill may not be considered a contaminant at another. Various contaminants include rubber bands, yellow post-it tabs, laser, fax, and photocopy papers, newspaper, magazines and cardboard. Paper clips and staples can be magnetically removed during the shredding process.

OWP has two distinct markets, which makes analyzing future markets extremely difficult. They are low quality OWP and high quality OWP. The low quality OWP is used in the production of paper board products and packaging materials. These products are quite often a mixture of high grades as well as other waste paper. The use of the low quality OWP helps to maintain the quality of the end product. The high quality OWP is used for printing and writing paper, tissue and market pulp. The remainder of this section discusses high quality OWP since low quality OWP can be introduced into many existing processing facilities.

In the past, high grade paper markets were established for pre-consumer white and colored ledger although a small amount of post-consumer scrap was also bought and sold. Prices for the different types of high grade waste paper vary greatly and range from $30-$200 per ton.

In the past several years, many voluntary office recycling programs have been established and are continually being established even though the value of the OWP is plummeting due to oversupply and the weak economy (Figure 5-4). These programs are not profit- orientated but are based upon cost avoidance. Many companies have discovered their refuse collection fees can be reduced by as much as 50 percent with an office recycling program in place.

The result of these collection programs has been a supply of OWP growing substantially faster then the demand. Currently markets for the post-consumer white or colored ledger are in the developmental stages. The inconsistent purchasing patterns of office white paper reflects the experimentation of many mills trying to utilize the abundance of OWP available.

The market for high quality OWP will remain fairly stable throughout 1993. Between 1993 and 1995, 12 large de-inking facilities and major expansions will nearly double the current capacity of high quality OWP. A recent study projected that the glut of recovered paper may get worse in 1994 before the market gradually moves into balance by 1995.

The market spotlight for the future use of OWP is primarily focused on printing and writing paper producers. Many users are demanding recycled content paper from these producers and are willing to pay extra to receive it. For some of these buyers, such as local, state, and federal agencies, the use of recycled printing and writing paper is required by executive order, law, and ordinance. Since federal agencies are larger consumers of office paper, many environmentalists have long argued that increased purchases of recycled paper would have a dramatic effect on the market for recycled paper. For example, it is estimated that the federal government purchases 300 million tons of paper per year, or about two percent of the national paper market. To help spur investment in recycling technologies and promote market opportunities for recycled paper, President Clinton signed an executive order in October 1993 requiring federal agencies and the military to buy paper containing 20 percent post-consumer recycled content by the end of 1994 and 30 percent post-consumer content by the end of 1998.

5.2.5 Residential Mixed Paper

Residential mixed paper (RMP) is a relative new paper category which covers a wide range of waste paper types and is not limited to grade or fiber content. RMP is composed of magazines, catalogs, direct mail and boxboard packaging (e.g., cereal and shoe boxes). Residential mixed paper generally is recovered from garbage, which makes it difficult to separate the contaminants. Collection of residential mixed paper would require a great deal of public education and can be very costly.

Mills that use residential mixed paper usually like to know the content of the mixed paper and generally deal with processors who consistently provide them with good mixed paper that will meet their specifications. The paper stock standards and practices circular PS-86 defines mixed paper as follows: "a mixture of various qualities of paper not limited as to type of packing or fiber content. Prohibitive outthrows may not exceed 10%." Metal, glass, plastic, and other non-paper material are considered contaminants. Mixed paper is used almost exclusively in recycled paperboard and construction grades, which include such products as roofing felts, door coverings, automotive fibers, and pipe coverings.

Markets for residential mixed paper in the U.S. and overseas are primarily the mills that produce recycled paperboard and construction grades. Exported wastepaper is consumed primarily by countries that do not have abundant supplies of virgin fiber materials but which have growing economies which have created a demand for paper and paperboard products. The largest importers of U.S. waste paper are Mexico, South Korea, Taiwan, Japan, Italy, and Venezuela. In many of these countries, recycling rates for waste paper are very high; because virgin fiber is not added, the quality of the recycled paper and paperboard decreases each time it is used. As an alternative, imported high-quality waste paper is used to help maintain the quality of the finished product.

Because the U.S. has a good supply of high-quality, inexpensive waste paper, the export market has continued to grow. Some export markets like India, with low labor costs have found it feasible to buy mixed paper and then separate it manually into specific grades for other uses.

In the past, residential mixed paper played an important role when demand for recycled paperboard products was high. However, as source-separated materials such as newspaper and corrugated, become increasingly available to the waste paper market, the use of mixed paper began to decline. Mixed paper value has always been less than source-separated materials like newspaper, office paper and corrugated. Currently some dealers are paying $5-10 per ton to export, while others are giving it to domestic mills.

The future for RMP looks bleak as long as there remains a glut of source-separated materials in the market. However, the future for magazines will increase in the next two years as new newsprint facilities which utilize new flotation de-inking technology come on line. This new de-inking technology requires 20 to 50 percent coated magazine stock to aid in de-inking and production of a stronger sheet. When the market price for corrugated increases, corrugated medium manufacturers and recycled boxboard manufacturers are likely to become interested in purchasing small quantities (10 to 20 percent) of the less expensive RMP.

5.2.6 Glass Containers

Glass, which has been in use for thousands of years, is a transparent substance, made primarily from sand, soda ash, and limestone. Glass containers are produced in three colors clear (flint), brown (amber), and green. Of these colors, flint has the largest number of applications and is usually in greatest demand by glass manufacturers. Brown or green glass is used in products where exposure to sunlight may cause the product to degrade.

The primary glass produced in the waste stream is the glass container, which is mainly composed of soda bottles, beer bottles, and condiment jars. Other glass products such as cooking ware, dishware, ceramics, windows and specialty glass also appear in the solid waste stream but are considered contaminants due to their chemical composition or heat resistant properties.

FIGURE 5-4

HISTORICAL PRICING TRENDS FOR HIGH GRADE OFFICE PAPER

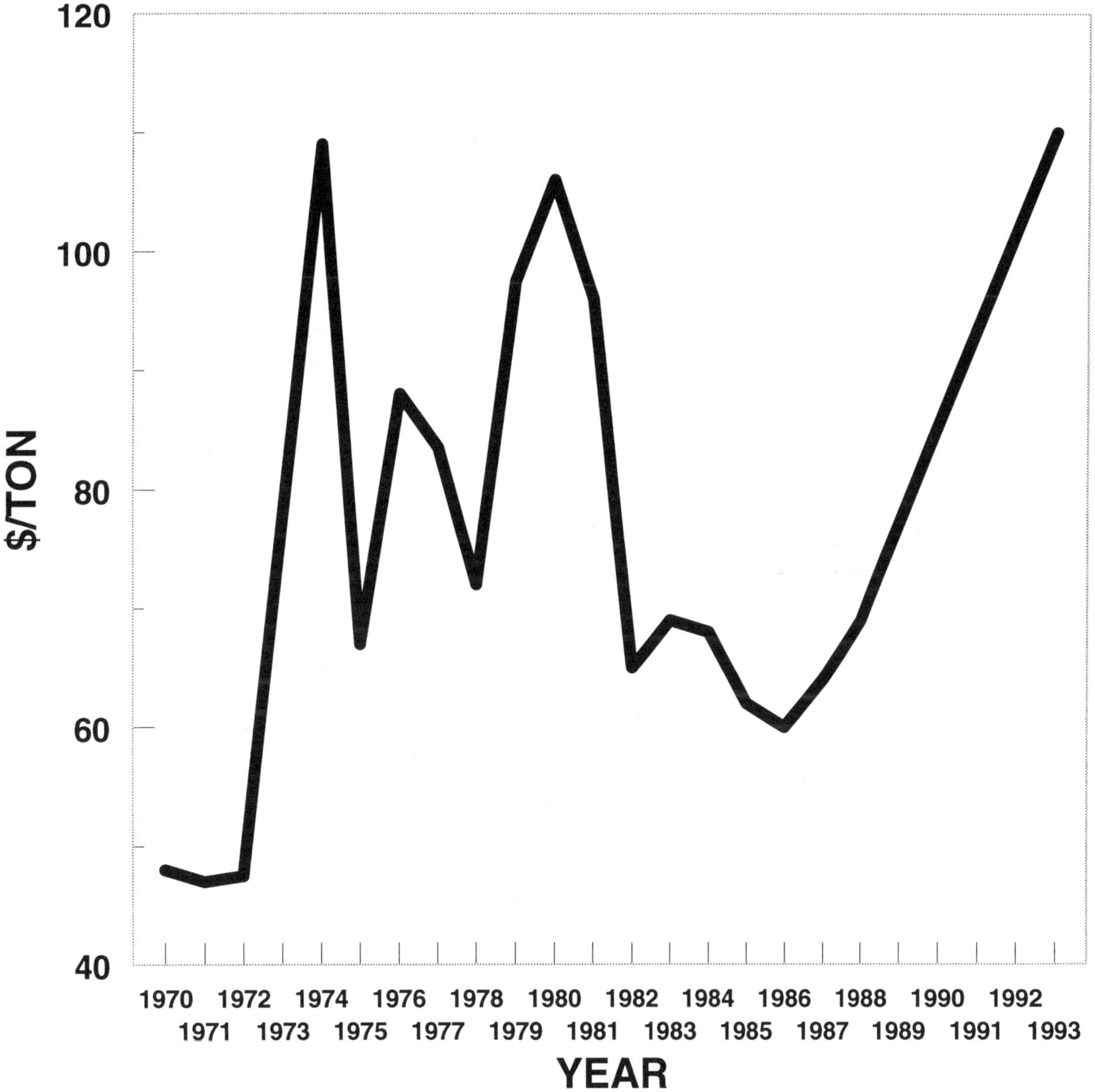

Source: Recycling Times, 1993
Reflects prices paid for white ledger

Most manufacturing facilities involved in glass recycling use only bottles and jars, i.e., container glass. These manufacturers also require collected glass to be separated by color, since the material is used to make glass of the same color. Mixing colors produces a low-quality glass container and in many cases an aesthetically unappealing end product.

If a manufacturer does not have appropriate processing equipment, the recycler (or middle processor) is required to remove metal, paper and other glass contaminants from the container glass as well as separate them according to color. Since glass furnaces operate at temperatures of 2,600°F, most metals will melt and corrode the furnace linings. Other metals such as aluminum form small balls that end up in finished products making them unusable. Melting mixed colors of glass and glass of varying chemical compositions in the same batch can lead to a foaming action in the furnace which produces off-color bottles with numerous air pockets within the glass. Ceramics and heat resistant glass do not melt at the temperatures used in a glass container furnace and show up in the end product as "stones" or other defects.

The primary markets for recyclable glass containers are the 75 glass container manufacturing plants in the United States. Other secondary markets include road construction, either on the surface called "glassphalt" or as a roadbase aggregate; filler aggregate in storm drain and French drain systems; the fiberglass industry; glass beads for reflective paints; abrasives; foam glass and other building materials.

In 1967, forty container glass manufacturers produced glass from 112 plants in 27 states. Today seventeen companies operate 75 facilities in 27 states. As glass containers lost market shares to cans and plastic bottles over the last 25 years, the glass container industry consolidated and reduced capacity.

In recent years, factors that have contributed to the increase in glass recycling are preservation of natural resources, reduction in litter, energy conservation reduced waste quantities, disposal cost, and reduction of raw material use. The natural resources in glass manufacturing are sand, limestone and soda ash. Although these resources are abundant in the United States, they are geographically separated by long distances, which leads to high transportation costs in procuring these raw materials. Thus, using recycled glass helps conserve oil and gas. The "bottle bill" legislation passed by many states in the 1970's encouraged glass manufacturers to use reclaimed ground glass called "cullet". Using cullet allows furnaces to operate at lower temperatures which extend furnace life, reduce energy costs, and lower stack emissions. The use of cullet in the manufacture of glass has increased steadily from 22 percent in 1988 to 31 percent in 1991.

Other considerations influencing private and government involvement in glass recycling are limited landfill space and increasing costs of waste disposal. By promoting recycling, the glass industry is also better able to maintain its share of the container market which is being impacted by the aluminum and plastic container industries.

The price paid for glass containers is determined by color, quality, and the extent to which it has been prepared, (i.e. crushed or whole). The prices paid for glass containers vary greatly depending upon proximity to glass manufacturing facilities. In some locations, such as upstate New York, some collectors are paying $5 to $8 per ton to market their green glass. The unstable market for green glass cullet is in large part due to the manufacturers. The quantity of imported foreign liquors bottled in green glass exceeds the production capacity for green color containers. In the United States, green glass bottle production represents approximately 13 percent of dedicated furnace capacity, while green glass containers compose 23 percent of the glass container waste stream.

During the 1980's, glass prices remained relatively stable for all colors of glass, at approximately $30 per ton. However, in the past several years, community recycling programs, mandatory recycling goals, third party processors and the increased sales of imported liquors have driven glass container prices to below $10 per ton in many areas of the country. Figure 5-5 illustrates the pricing trends for recyclable glass over the last 11 years.

Probably the greatest influence on cullet and bottle prices in the late 80's was the supply of new material from communities with mandatory recycling programs. As more communities implemented recycling programs to extend the life of existing landfills, a new flood of glass caused prices to decrease. Because mandatory recycling was motivated by cost-avoidance, communities were willing to give glass away, if necessary.

For decades, intermediate and secondary processors produced cullet from natural resources and reclaimed materials, which was then sold to glass manufacturing facilities. However, in the 1980's, due to the abundance of available container glass, many glass manufacturing plants spent millions of dollars on glass benefication facilities at their plants. The introduction of new buyers has helped to stabilize the market prices.

FIGURE 5-5

HISTORICAL PRICING TRENDS FOR GLASS CONTAINERS

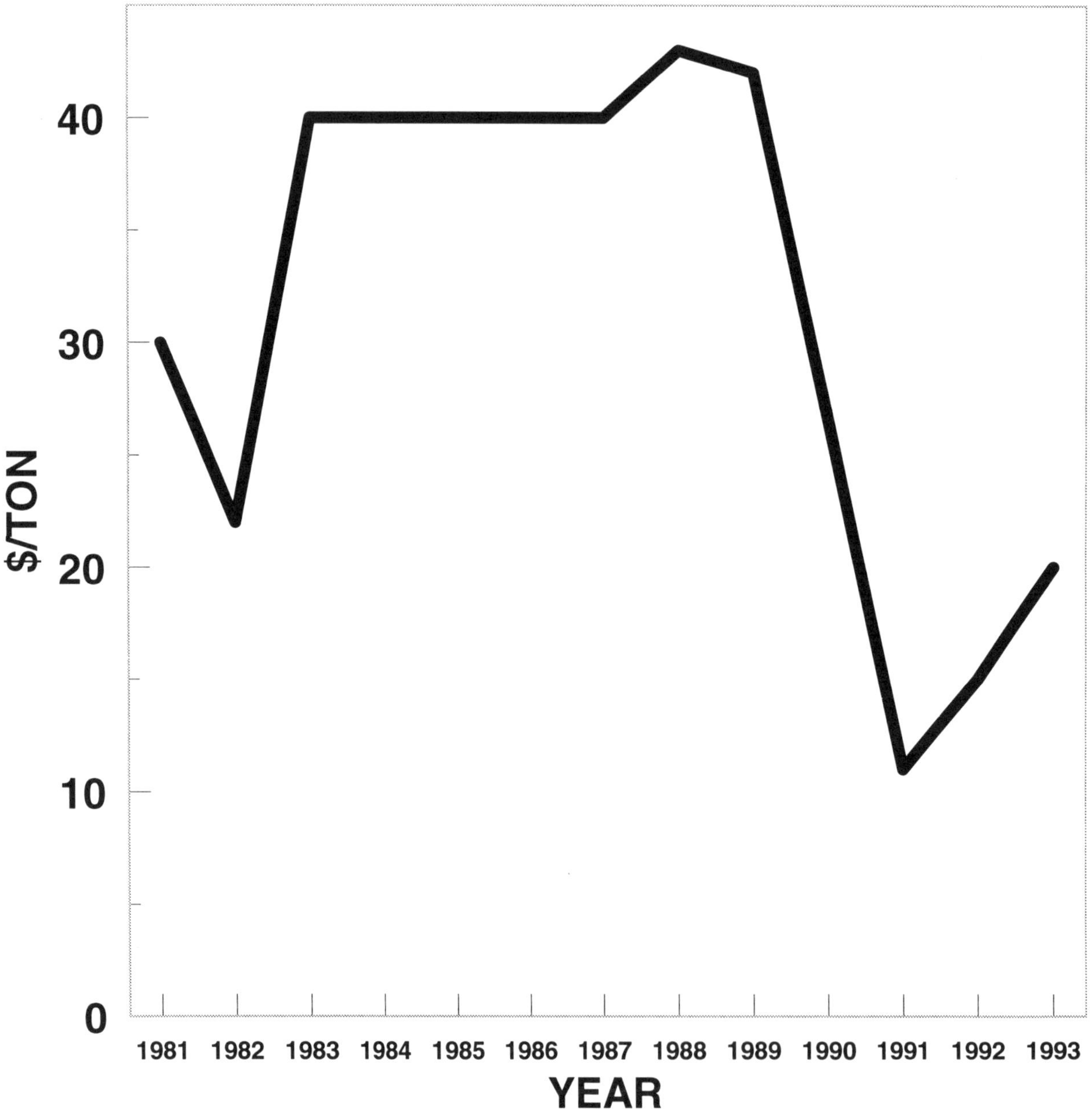

Source: Recycling Times, 1993
Trend reflects prices for clear (flint)

Last year, Owens-Brockway, the largest U.S. glass container manufacturer with 22 plants and a 30 percent share of the total container glass production, decided to cease the cullet processing and benefication activities in most of their facilities. Now, they can concentrate on producing glass containers from high quality material purchased from third-party processors instead of paying the high costs associated with cullet production. Thus, bottle manufacturers do not have to worry about varying cullet quality and can replenish stock piles with reliable, on- time deliveries.

During 1991 and early 1992, there has been a growing interest in third-party arrangements between processors and manufacturers. As a result, existing processors have expanded their facilities and new processors have entered the market. Due to the limited capacity and small limited competition between processors in a given region, the new third party arrangements have drastically lowered the price of recycled container glass.

With cullet prices heading downward in most parts of the country, many communities are contemplating removing glass containers from their recycling programs. Communities have discovered that glass recycling is a labor intensive and time consuming endeavor which can prove to be expensive with falling cullet prices. In many cases, however, the public relations benefits and avoided tipping fees are felt to outweigh the collection and processing costs.

The future market value for recyclable container glass is dependent on three issues: new processing strategies, developing new markets, and recycled content legislation. Manufacturing companies are currently experimenting with increased levels of reclaimed cullet and utilizing mixed color cullet in batches to ease the saturated markets. In the first quarter of 1991, Owens-Brockway reported green cullet in its batches at almost 60 percent, up from 1990's value of 47 percent. They have also reported batches of green container glass at levels approaching 80 percent. This demand is above post-consumer green cullet availability.

Another strategy has been to incorporate green cullet and mixed-color cullet into amber furnaces. Coors Glass Manufacturing is currently using three-color-mixed cullet with a total cullet to batch ratio of 33 percent and feel that a 50 percent ratio can be achieved.

The challenge in developing new markets for reclaimed glass is finding higher value uses so the glass does not compete with low value aggregate or sand. With this in mind, fiberglass, glass bead, and abrasive markets offer the greatest potential. Glassphalt and other construction applications are slow in developing potential.

Current legislation in Oregon and California requires glass container manufacturers to use minimum quantities of post-consumer waste glass in the production of new glass containers. Oregon's law requires container manufacturers to use 35 percent by 1995, while California requires a gradual increase from 15 percent in 1992 to 65 percent by 2005. California has also set cullet content requirements of 10,20, and 30- percent which are to be met by the fiberglass insulation manufacturers by 1992, 1994, and 1995 respectively. The Mineral Insulation Manufacturers Association supports the legislation because it gives manufacturers the right to refuse cullet which does not meet the cullet specifications outlined in the legislation.

5.2.7 Scrap Metals

Two types of metals are commonly recycled, ferrous metals and nonferrous metals. The nonferrous metals include such materials as aluminum, brass, copper, lead, and zinc. By far, the most common recyclable of this group is the aluminum beverage can, which is discussed separately in the next section. Ferrous metals include cast iron, "tin cans", and other steel products and stainless, steel, including industrial scrap, car bodies, and household appliances.

Scrapped autos are a major source of ferrous scrap for the secondary metals market. On the average, an automobile remains on the road for about 10 years. This time is increasing because of special anti-corrosive treatment of metals and greater precision and longevity of the power train. Studies done in the early 1980's of all deregistered cars indicate that 60% are processed by scrap shredders and then sold as ferrous scrap to iron and steel mills. Another portion of deregistered autos is stocked by the auto dismantling industry as an inventory of used parts.

A substantial portion of the total tonnage of shredded ferrous scrap is exported. According to the Institute of Scrap Recycling, carbon steel scrap exports totaled 9.49 million tons in 1988 (a 5.5% decrease from 1987) while stainless steel exports increased 44% from 1987 to over 243,300 tons and alloy scrap jumped 161% to over 362,700 tons in 1988.

Since ferrous scrap from junked autos and other industrial scrap represent a significant portion of ferrous that is already shipped to the steel industry, this section focuses on another major source of ferrous scrap, the ferrous can. In the residential waste stream, the tin-plated steel food can is the largest volume ferrous metal product discarded.

Ferrous cans are made up of three general types of containers. The predominant type, in terms of quantity, is the tin-plated steel food container that is coated on the inside with a thin film of tin to preserve freshness of food.

The second type is the steel can that does not have the tin plating. The third type is the bimetal can, used primarily for beverages, that has a steel bottom and sides and an aluminum top with a pull tab.

The steel industry prefers the tin-plated steel container after the tin and other contaminants are removed in a detinning facility. In some cases, the industry will accept small quantities either of tin-plated or bimetal cans directly from scrap dealers or municipal recycling programs. These containers should be baled or densified. Detinners (companies that remove the tin) prefer the tin cans free of labels, thoroughly rinsed out, and shredded in one-to-two inch square of thin plate. In some cases, the containers can simply be flattened and shipped loose with labels attached.

The only domestic source of tin is in the millions of tons of post-consumer tin-plated steel food cans, which has given rise to the detinning industry. After the tin is removed by dipping the container into a chemical solution that separates the tin from the steel, it is reused for its original purpose, tin-plating. About six pounds of tin is recovered from a ton a scrap cans. It is important to note that the detinning industry serves the purpose of upgrading the quality of the steel by removing tin and other contaminants. The resultant steel is a fairly high and uniform grade that can be used in electric arc furnaces (mini-mills) or the traditional basic oven furnace to produce any type of steel desired. Some tin-plated steel is purchased directly from suppliers without the tin removed, but it can only be used in very small quantities.

The detinning industry views the increased interest in recycled tin cans as a potential source of additional tin. An industry representative indicated there was an opportunity for a fivefold increase. As more recycling programs include this material, additional detinning plants may go on line. The potential exists for continued and expanded recycling of tin-plated steel cans.

Bimetal cans pose a far more difficult problem; until the industry finds an economical way to separate aluminum from steel, the potential recycling of these containers is very limited. This limitation may be offset by the current trend of reduced demand by the beverage industry in using this type of container because it is not easy to recycle. This has also resulted in the sudden shift to the all-aluminum can in certain markets.

The backbone of the source separated, tin can recycling market was the detinning industry. However, the number of detinning plants fell from 15 in 1968 to seven in 1987 as a result of competition in the beverage container market from plastic and aluminum. Sufficient capacity exists in the detinning industry to accept additional tin cans. However, it is difficult to economically collect and process tin cans. As the cost of solid waste disposal increases, solid waste officials are becoming more interested in removing tin cans from the waste stream.

In the past, the steel industry has preferred to buy the higher grade steel from the detinning industry after the tin has been removed. Tin is a contaminant and gives steel a flaky or cobbled appearance. However, recently steelmakers are taking notice of the scrap metal can market for a variety of reasons. Steelmakers efforts to regain a larger share of the beer and soft drink can markets include more recycling of bimetal beverage containers (steel body with aluminum lid). Purchase of these scrap cans demonstrates the industry's action to effect this recycling as well as reduce litter and lessen the requirements for landfilling. Steel mills have overcome such problems as the need to minimize residual tin. The steel can, among the lowest cost source of ferrous scrap, is desirable scrap if it is relatively free of contaminants and if its content is limited properly in the scrap mix. Lastly, a growing amount of scrap is available from source separation operations and other sources, such as waste-to-energy plants that take in municipal solid waste and magnetically separate steel cans and other ferrous materials for resale to scrap markets. This relatively stable trend is presented in Figure 5-6.

Steel producers lead the current market for recyclable tin cans, although many scrap processors figure the detinning industry is where market growth will occur. The steel industry is in an economic slump brought about by the recession. Currently, steel mills are running at three-quarters capacity and will receive no immediate help from the appliance producers and automobile makers. In 1991, 5 percent fewer appliances were produced and 11.5 percent fewer automobiles were sold than were produced/sold in 1990. This decrease in demand combined with automakers replacing steel with plastics has the steel industry very concerned. It is anticipated that future markets for tin can scrap metals will increase when new detinning facilities are constructed with upsurge in the economy.

5.2.8 Aluminum

Recycled aluminum has long been the highest-valued commodity of all the secondary materials. This is due largely to the high cost of mining and processing virgin material, the large investments in capital equipment, and high amounts of energy needed to smelt primary aluminum. With the high costs in processing aluminum and demand for all-aluminum cans growing, recycling has helped aluminum manufacturers remain competitive with other forms of beverage packaging. The 95% reduction in energy consumption to recycle aluminum cans coupled with a highly competitive foreign bauxite market makes recycling the most viable alternative for the domestic aluminum industry.

FIGURE 5-6

HISTORICAL PRICING TRENDS FOR SCRAP METAL

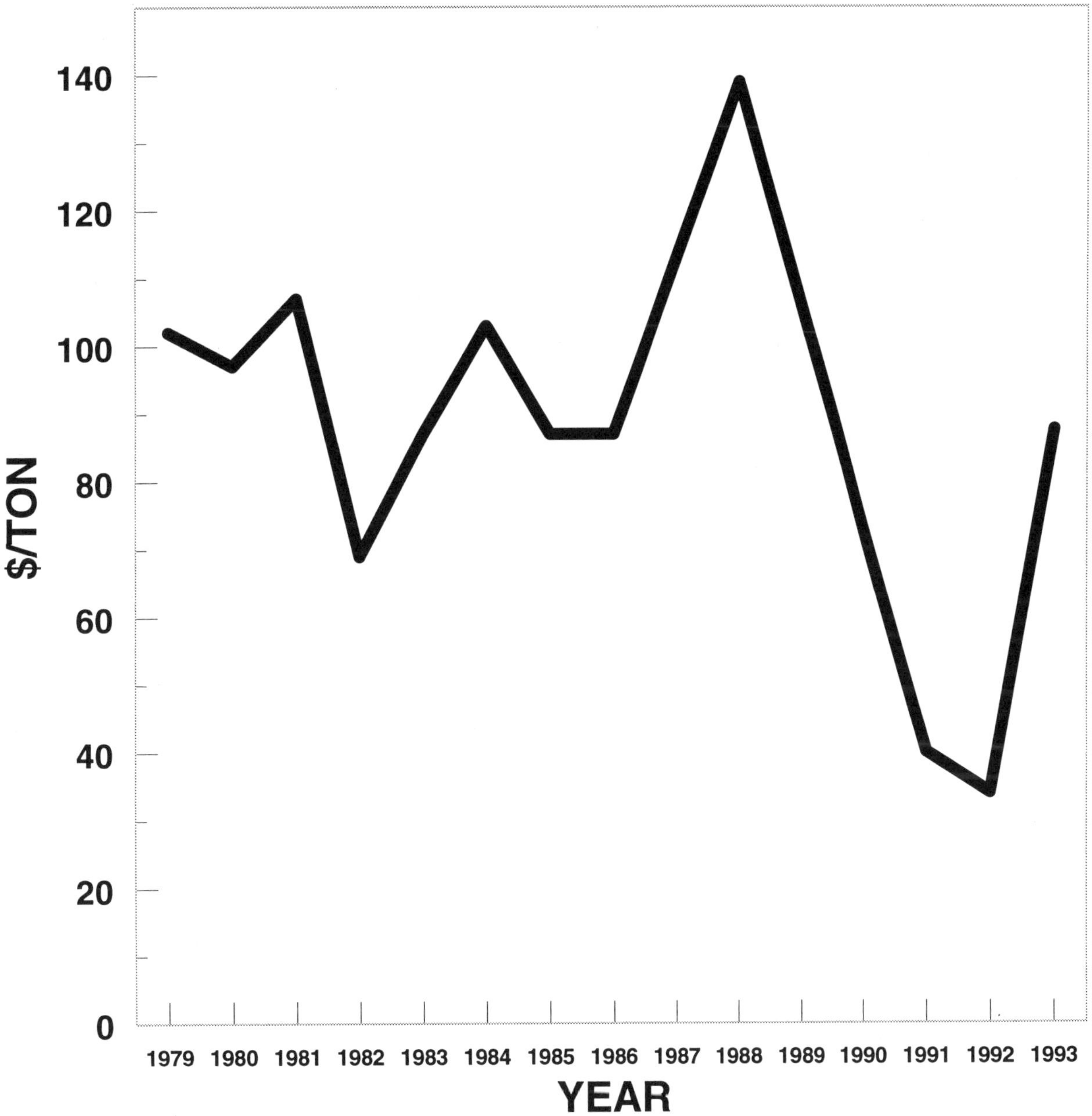

Source: American Metal Market/Metal Working News, Recycling Times, 1993
Trend reflects prices paid by end users

Bauxite, the principal raw material from which aluminum is made, is one of the most abundant components of the earth's crust, and the technology to mine it has been known for approximately 100 years.

Domestically, bauxite deposits are very limited with the U.S. aluminum industry depending on foreign sources for most of its supply. The high costs of mining and processing aluminum ores caused the industry to realize that scrap aluminum had value as a resource and that it could be collected and remelted for use in new products. The technology for recycling scrap aluminum was developed about 75 years ago, but did not include recycling cans, which is a relatively new development brought about primarily by trends in the beverage packaging industry.

Over 200 registered aluminum alloys, each having different mixtures of components, are broadly classified as wrought and cast aluminum. Wrought aluminum (e.g., beverage cans), is used primarily to produce beverage containers and foils and is collected for reuse in like products. Cast aluminum consists of scrap castings that are reprocessed into ingots of casting alloy. Both components of aluminum in the waste stream are reviewed below for their recyclability.

5.2.9 Aluminum Cans

The aluminum can, commonly called UBC (used beverage can) in the aluminum industry, continues to dominate the beverage can packaging market with an average share of over 95%. UBC's are often collected, baled and shipped to processors however, two of the largest recyclers of aluminum beverage cans, ALCOA and Reynolds Aluminum, prefer to receive their used beverage cans (UBC) loose and flattened in tractor trailers so they can examine them for contamination. Contamination is anything that is not an aluminum beverage can: dirt and moisture on cans, plastic, steel, iron, lead, paper, or other common items. In cases where large quantities of aluminum will be recovered, processors often provide the communities or organizations with the necessary processing equipment (i.e., flattener, blower, and tractor trailers). Recycling plants may return highly contaminated shipments at the shipper's expense.

UBC's are currently collected with other types of aluminum (small quantities of aluminum foils are acceptable) and are used mainly in the manufacturing of new aluminum cans. Once collected, the UBC's are shipped to conversion mills to be remelted and formed into high quality aluminum sheet metal. These sheets are then sold to can manufacturers to produce new aluminum beverage containers. Recent recycling rates for UBC's (68%) indicate that a single can may be recycled as many as six times in one year. The aluminum industry has encouraged and promoted recycling by developing aluminum buy back centers and providing UBC processing equipment and transportation to companies, groups and individuals interested in recycling.

When the all-aluminum can was first introduced into the packaging market in 1963, it capitalized on the knowledge that recycling aluminum only takes 5% of the energy that producing virgin material from bauxite does. Since the bauxite supply in the U.S. was and still remains very limited, recycling the aluminum can made it possible for domestic aluminum companies to decrease dependence on foreign bauxite supplies and to help conserve our own natural resources. In the 1970's many states started anti-litter campaigns and passed legislation requiring consumers to pay deposits on beverage containers. This increased the recovery rate of UBC's and helped to make recycling of UBC's possible.

Historically, the UBC market has been dominated by the large aluminum processors such as Alcan, ALCOA and Reynolds which have been able to minimize the competition for UBCs from secondary smelters and export markets. However, in the past few years, these previously excluded industries have begun to purchase UBCs from smaller sources such as small communities and individuals. The involvement of the secondary smelters and export markets has had a dramatic effect on the market price for UBCs, driving the price from an industry high of 76 cents per pound in early 1989 to the current price of 25-40 cents per pound. Figure 5-7 presents the historical pricing trend for UBC.

In 1990, approximately 1.9 billion pounds of UBCs were collected and processed up from 0.6 billion pounds in 1980. Another billion pounds of scrap generated at can plants was also processed bringing the recovered UBC and scrap recovery to 2.9 billion pounds. Recent surveys indicate that the aluminum industry has the processing capacity to melt in excess of 3 billion pounds annually. This near capacity situation held market prices steady throughout most of 1990.

The mandatory recycling programs established in 1991 and 1992 saturated the aluminum market with UBCs. The aluminum industry which was operating at near capacity levels could not purchase the excess aluminum and as a result aluminum market prices dropped. Aluminum is now selling at near record-low prices due to the following reasons. Since 1990 producers from the Commonwealth of Independent States have pumped as much as one million metric tons of metal into the Western market per year (a metric ton equals 2,046.2 pounds) as a means of raising badly needed cash with limited domestic aviation or defense industry demands. Other producers played a game of chicken, daring their competitors to be the first one to close production plants. With stagnant consumption of the light metal in the current recessionary marketplace, the natural result was plentiful aluminum at cheap prices, about

49 cents per pound, an all-time low considering inflation. Warehouse stocks of aluminum are projected to exceed 2,000,000 metric tons. It is in this weak market that scrap processors tried to sell recyclable aluminum with prices continuing to fall.

The price for UBC's can vary from day to day, based on the price for primary aluminum, the quality of the collected UBC, and the supply and demand of UBC's on the market. Due to these factors, it is impossible to get accurate past prices for the UBC. Currently, prices are at 25-40 cents per pound, depending on the quantity, quality, and preparation of the material.

The same factors that influenced prices in the past will continue to have control on future pricing but the implementation of municipal, county, and statewide recycling will also continue to have a dramatic effect. In the past, UBC recycling was done predominantly by the private sector and financed by the aluminum recycling industry. However, mandatory recycling, unlike private recycling, is based on preservation of landfill space and cost avoidance associated with landfill costs. With the increasing supply of UBC on the market and fewer profit-motivated parties involved, the aluminum recycling industry may be able to reduce prices while receiving increasing supplies of UBC. This, however, will not cause an oversupply of aluminum cans. Theoretically, the aluminum industry could recycle every aluminum can made provided it is prepared properly.

5.2.10 Other Aluminum

In addition to UBCs, municipal solid waste also includes other aluminum products (e.g., aluminum foil, lawn furniture, window frames, outdoor play sets, and automobile parts). Except for UBCs and aluminum foil (which are wrought aluminum), the above mentioned products are cast aluminum. The difference between wrought and cast aluminum is primarily how it is manufactured into finished products. Wrought products, like aluminum cans and foil, are stretched and pressed by a mechanical process to form the product. Cast aluminum is melted and poured into forms or casts and cooled to harden into shape.

Since the can recycling industry deals primarily with UBC, the secondary aluminum sectors deal with many of the remaining 200 or more alloys. Aluminum by itself does not provide the strength needed for the uses stated above. Other metals are therefore blended in to form an aluminum alloy. The chemical composition of each alloy is different to provide the specific characteristics needed for each product. For instance, aluminum is mixed with iron and manganese and is used in automotive parts to help reduce weight in cars, while providing needed strength and rigidity.

FIGURE 5-7

HISTORICAL PRICING TRENDS FOR ALUMINUM UBC

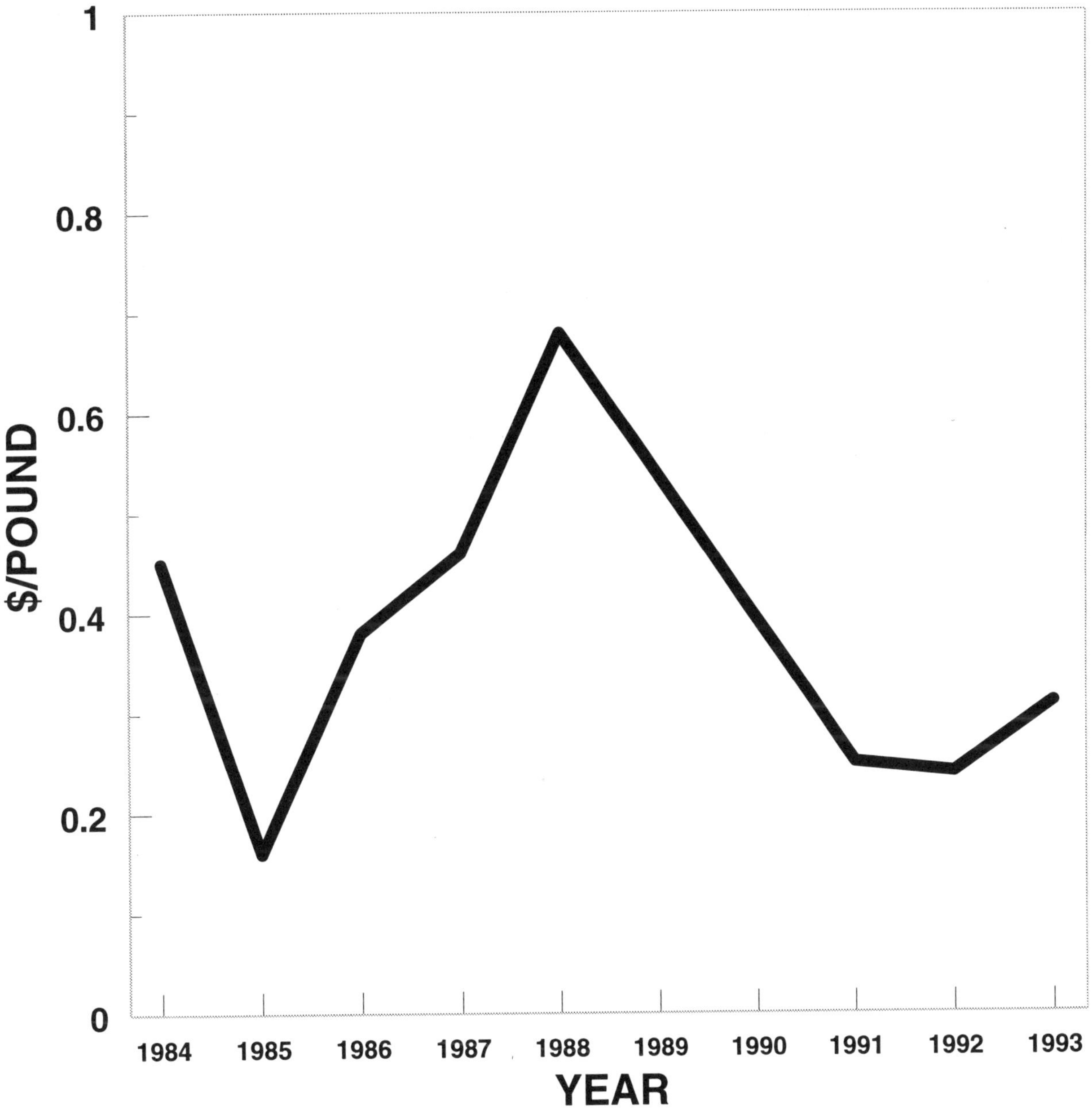

Source: Recycling Times, 1993
Trend reflects prices paid by processors

FIGURE 5-8

HISTORICAL PRICING TRENDS FOR HDPE PLASTIC CONTAINERS

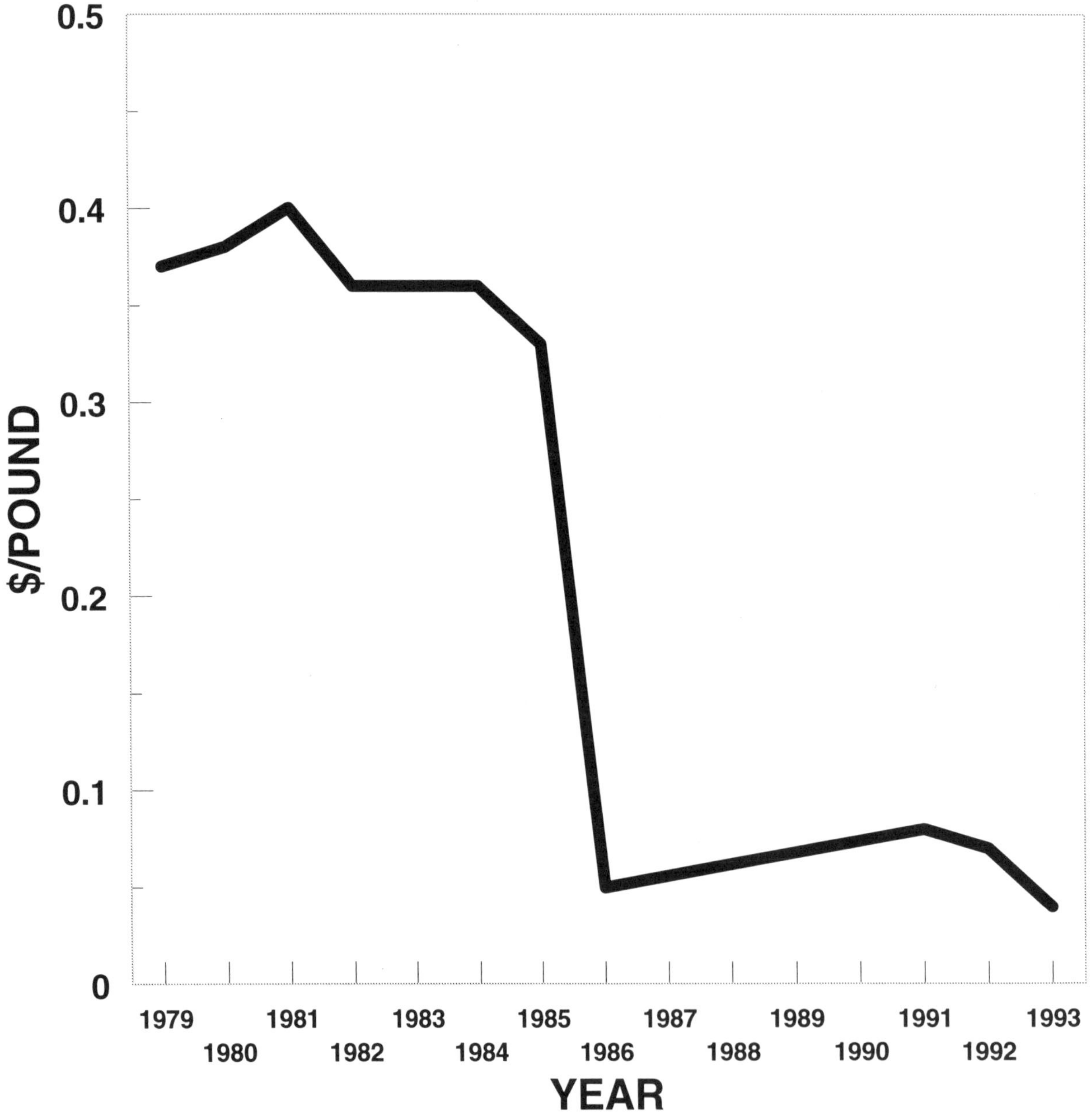

Source: Recycling Times, 1993
Trend reflects natural HDPE paid by processors

Aluminum foil is almost pure aluminum. Since it is free of most alloying metals processors primarily use it as a "sweetener" in their scrap furnaces. The influx of pure aluminum into the scrap aluminum batches "dilutes" the metal alloys and "aids" in the cleansing process.

Industries and households sell their aluminum to a scrap dealer, who generally separates it into different grades for sale to the secondary aluminum industry. Sorting is usually done by scrap dealers who know the industry and the type of alloys. They buy most kinds of ferrous and nonferrous metals, separate and upgrade them, and then sell them to their prospective markets. Unlike many recyclables that go through little processing before being returned as finished products, secondary aluminum requires as many as four processes before it is completely recycled.

A large number of products are produced from the various alloys of aluminum. The secondary aluminum industry produces different grades of casting and extrusion ingots that are sold to foundries and die casters for uses in many products. One of the largest buyers of secondary aluminum is the auto industry. Transmissions, engine blocks, brake components, pistons, and other automotive parts are all cast from aluminum or aluminum alloys. The demand for specific grades of secondary aluminum is determined by the properties offered by various alloy chemical compositions. Demand for recycled or raw scrap aluminum is growing in foreign countries such as Japan, Korea, and Taiwan which are known for their automobile production.

The value of remelting secondary aluminum into ingots for new products was realized in the early 1900's by the secondary aluminum industry. This industry was the primary market for scrap aluminum until the last 20 years. In the late 1960's, demand for secondary aluminum alloys grew due to the increased uses of aluminum in automobiles. Demand continued into the 1970's, and new competition entered the secondary market from primary producers adding equipment that used secondary scrap as well. Foundries and fabricators also added scrap aluminum capacity.

As the U.S. economy strengthened in the 1980's, the dollar value overseas made it cheaper to import aluminum ingots than to use the domestic supply, causing a major fall out of secondary aluminum recyclers. In the early 1980's, there were 65 companies in the secondary aluminum industry; today there are only 40 companies left.

Primary aluminum prices have historically influenced aluminum scrap prices on a day-to-day basis and will continue to play a major role in the very volatile pricing of aluminum. The recent weak dollar overseas has stimulated countries like Japan to import U.S. secondary aluminum for its automotive needs.

The main factor affecting other aluminum scrap prices is the United States and world economies. Currently, the United States and many European countries are in the middle of recessions. These recessions influence the monies people have, or are willing, to spend on items such as homes and automobiles which are large markets for the aluminum industry. Other products such as lawn furniture and outdoor play sets have also seen reduced sales.

On the other hand, automotive design trends imposed by federally mandated mileage standards have increased the use of aluminum per passenger car by some 30 percent over the past 10 years. Some estimates say that the 1995 usage per car will exceed 200 pounds and that 500 pounds is a possibility by the year 2000. Current news reports indicate that the production of aluminum automobile frames will start shortly.

In 1991, Reynolds has funded a major advertising and public education campaign in its five test markets to educate consumers about the importance and ease of aluminum foil recycling. In 1992, Reynolds plans to expand this program to half of its national market.

5.2.11 Plastics

Plastic resins are synthetic materials made from oil and natural gas that are combined in a polymerizing process. Each resin has a different molecular structure that gives the material unique qualities and its value as a material. Today, more than 200 different types of resins are used to produce plastic products within the United States. This large number of resins had hindered the plastic recycling effort due to identification problems, until the Society of Plastics announced its "voluntary coding system" (which appeared on the bottom of many plastic containers) in 1979. Plastic packaging material such as containers, films, and wrappers provide several benefits which have broadened their use. These include breakage and leakage resistance, hot/cold insulation properties, versatility in shapes and sizes, and reusability, (e.g., using grocery bags or bottles for other purposes). However, the durability which has given plastic its advantage is also a disadvantage. when disposed, much of this packaging is not biodegradable and persists in the environment for a long time.

The primary types of plastic resins used in containers including the following:

♲ 1	"PET"	polyethylene terephthalate
♲ 2	"HDPE"	high -density polyethylene
♲ 3	"PVC"	polyvinyl chloride

4	"PP"	polypropylene
5	"LDPE"	low-density polyethylene

These five plastic resins compose roughly 98 percent of the plastic containers manufactured in the United States today. Plastic containers tend to be made from these readily identifiable resins and represent approximately 49% of the plastic packaging. Film plastic, such as plastic bags, accounts or 37%, while coatings and closures are 9% and 5%, respectively.

PET is a strong but lightweight form of polyester used for soft drink bottles, liquor bottles and other food and non-food containers. Recycled PET is used to make soft drink bottles, other containers, fiberfill in jackets and sleeping bags, carpet fibers, industrial strapping and may other consumer items. In 1990, plastic soft-drink bottles were recycled at rates exceeding 31% and resulted in 250 million pounds of PET and 127 million pounds of HDPE from the base caps. However, market research studies conducted by the University of Toledo for the Plastics Recycling Foundation suggested that a potential market existed for 700 million pounds of PET from recycled soft drink bottles in non-food applications. Thus, in the opinion of the plastics industry, the market is far from being saturated.

There are two types of HDPE bottle grade material, they are the homo-polymer and the co-polymer. The homo-polymer HDPE (blow molded) has a stiffer molecular structure and is used for dairy, water and juice bottles. The co-polymer HDPE (injection molded) has a more flexible molecular structure and is chemically more resistant to bottle contents such as detergents and household cleaners. These two types of HDPE are incompatible and cannot be mixed together.

PVC is known for its flexibility, high chemical resistance, and lower cost. It is mainly used in flexible bags and piping materials, however it has also been used for food jars and bottles. PVC has limited thermal stability and tends to degrade quickly when processed.

These three types of resins represent nearly 94% of the plastics market in the United States. HDPE and PET are included in most recycling programs while PVC is usually excluded due to its low market share.

The best markets for plastics currently are for separated plastic compounds. Therefore, easily identifiable and separated plastics are most readily marketable. PET soda bottles, and HDPE milk, juice and water bottles are in this category and are the most commonly recycled post-consumer plastics.

The fact that plastic containers are a high volume, low weight material affects the costs of collection from the community, processing and shipment to market. Collected plastics require manual sorting by trained people, by material type and color. Many plastic processing centers are hesitant to purchase post-consumer PET from programs where sorting was carried out buy non-trained personnel for fear of PVC contamination. It takes only a few PVC bottles to contaminate a 40,000 pound truckload of PET.

Baling the sorted material is one approach to reducing the volume. A more effective approach is grinding the material in a granulator. This, in particular, increases the importance of enforcing strict separation procedures. A careful assessment of the volume of plastic containers generated and collected from a community will influence processing, along with location of the buyer. The primary buyers of plastic waste from municipalities are few, and volume reduction through baling or grinding is essential to reduce transportation costs.

Since World War II, plastics have been used in an increasing number of products due to their versatility, durability, and lightweight characteristics. The introduction of the plastic (PET) soda bottle in 1979 caused a phenomenal amount of growth within the plastic packaging industry.

The history of the plastics recycling industry has been primarily propelled and shaped by three main factors; deposit and redemption mechanisms, mandatory state recycling goals, and the public relations factor due to the recyclability of aluminum and glass beverage containers. In the early to mid 1980's, several states enacted "bottle bills" which first focused attention on post consumer plastic recycling. The material returned from these programs was extremely contaminant free. This facilitated the marketing of the material. quantities recovered by these programs was very low which increased the expense of transportation.

Recent state mandated recycling goals have led to numerous recycling programs across the nation. These programs have produced a glut of reusable PET and HDPE plastics available to the plastic resin markets. As a result, prices for recyclable plastics have fallen over the past several years (Figure 5-8). Some communities are fortunate enough to live in close proximity to plastic resin producers thereby receiving some monetary value for their collected plastics. Many are forced to dispose of their collected plastics in landfills because it is less expensive then paying the transportation costs.

For the past several years, recycled PET has been used in the manufacture of new bottles. Before March 1991, these bottles were limited to non-food and non-beverage applications (fiberfill, strapping, carpeting, engineering plastics, and geotextiles). Thus, while no federal law specifically restricts the use of recycled plastic in food and beverage packaging, packagers and resin producers concluded that cleaned post consumer scrap converted into pellets and mixed with virgin materials cannot be assured to be contamination free. At this same time, market pressures were being asserted on soft drink giants to get a recycled content bottle on the store shelves next to the recycled aluminum, steel, and glass containers. As a result, chemical recycling or regeneration was born, and shortly thereafter Coke and Pepsi began using recycled content bottles.

Regeneration is the process of breaking down plastics into its two monomers (dimethyl terephthalate, DMT and ethylene glycol, EG) by adding chemicals and heat. The DMT is then purified to produce near virgin quality PET resin. Regeneration processes are costly requiring tremendous amounts of energy and very clean scrap plastics. Regenerated PET bottle resin has been estimated to cost 10-50 percent more than a virgin PET bottle. This cost factor appears to be the reason why no one is yet producing a new bottle with greater than 25 percent regenerated resin. Regeneration has many critics stating that plastics regeneration is merely a public relations effort by the soft-drink industry to introduce a recycled content bottle to the public.

Regeneration accepts only extremely clean scrap and is too costly to produce bottles with more than 25 percent regenerated resin. Regeneration of plastics might actually hinder future recycling efforts by keeping market prices at low levels. These low prices may discourage cities from collecting more plastics until they know there is an established market. At the same time, processors are reluctant to invest in equipment without an assured flow of quality materials. Further, another threat to development of this market is the refillable PET bottle which can be cleaned and refilled up to 25 times before being recycled. This new container is now used in several counties and may be introduced in the United States in the next few years.

Notwithstanding these problems, plastic manufacturers have come under increasing pressure in recent years to use the recovered materials from community-based plastic recycling programs. The industry has funded an extensive research program nationally to study collection and sorting technology at the Center for Plastics Recycling Research at Rutgers University - The State University of New Jersey. As an outgrowth of this research at Rutgers a number of resin reclamation technologies have been developed and offered for commercialization.

Further, three of the nation's largest plastics resin producers, Phillips Petroleum, Quantum Chemical, and Union Carbide each opened major recycling plants. The combined capacity of these new facilities is anticipated to increase the current demand for scrap plastics by another 100 million pounds annually. In addition, nearly every major company producing consumer products using plastic bottles has announced plans within the current year to use recycled plastic bottles.

The production of products from recycled commingled plastic materials is a small but growing industry. Developing markets for first-ever products to replace wood, concrete, and steel with plastics has been an inherently slow process, but has been stimulated with new specialized extruder/molder technology developed at Rutgers in recent years. Several manufacturers have reported plans for increased production capacity to produce such products as plastic lumber, picnic tables, landscaping timbers, fencing, pallets, car stops, and park benches.

5.2.12 Yard Waste

Yard waste is one of the largest contributing components of the municipal solid waste stream. By weight and volume, it is only exceeded by paper products. Nationally, it represents 15-20% of the solid waste stream. By definition, this waste is generated from the homeowner's yard, tree and brush trimmings, garden waste, grass clippings, and leaves. The majority of this material is generated on a seasonal or cyclical basis. Tree trimmings enter the waste stream normally in the late winter or early spring, while garden waste and grass clippings are produced during the spring and summer months. Yard waste is also generated by landscape maintenance firms from the upkeep of both residential and commercial properties.

For yard waste to be recycled from the homeowner, source separation is an absolute requirement. Material is best prepared at the source. Tree or brush clippings should be bundled, garden waste should be put in containers, and leaves put at curbside in bags or loose for vacuum pickup. Bagging leaves, garden waste, or grass clippings creates problems for the composting process with an additional processing step required to remove the bags. Plastic bags are a contaminant and interfere with the decomposition process of composting. Furthermore, residual pieces of bags in the finished compost product are aesthetically undesirable and lowers the market value.

Some of the materials would have to be processed prior to composing. Brush, woody materials, and garden waste need to be ground to provide proper size and surface area for microorganisms to work on. Leaves generally do not need to be ground, although grinding does accelerate the digestion process.

The market for organic compost is not as developed as that of the commodity recyclable materials. Typical markets for compost are greenhouses, nurseries, golf courses, landscape contractors, turfgrass farms, cemeteries, agriculture, and topsoil suppliers. It is also used as landfill cover, along roadsides for erosion control, and for other non-food horticultural activities. Most often, yard waste composting projects realize minimal to zero revenues, claiming benefit from the avoided cost of landfilling if the product is taken away by an end-user or from the avoided cost of topsoil if the product is used as landfill cover.

Markets for compost can be divided into four major categories which include growers, processors, wholesalers and bulk users. Growers can include golf courses, nurseries, landscape companies, lawn maintenance businesses and sod and sod service businesses. Processors refine the compost to meet specific conditions and market the material. Processors may include topsoil companies, fertilizer companies or sand and gravel companies. Wholesalers typically purchase and sell bagged compost at garden supply or nursery type businesses. Bulk users utilize large quantities of compost to cover acreage for land reclamation, landfill cover or for park or roadside stabilization after construction. Since most road construction is done by state DOT's, discussions with DOT officials could result in a significant market.

-6-

PUBLIC EDUCATION PROGRAMS

Public education is one of the most important and difficult to manage elements of any recycling or waste management program. Without the public's cooperation and continued support, a local recycling program will never reach its maximum potential. Research indicates that, unless people learn to participate in recycling efforts on a natural or habitual basis, participation rates will fade over time. In this section, we will discuss some basic theories that should influence planning of public education initiatives. We will offer some do's and don'ts of persuasive communications and talk about the development and implementation of recycling oriented programs. We will conclude the section with some examples of some of the most successful public education strategies and information as to additional sources of information and public education resources.

6.1 How Do People Learn

Why should it matter? Given the rarity of public education funds, efficiency and rate of return on each dollar invested is critical. The most effective programs are responsive to human psychological needs for knowledge development. Before we begin to plan for recycling education, it is important to understand how people accumulate knowledge or form patterns of behavior. While this book was not intended to serve as a psychology primer, it can suggest key points, that if understood, could influence the direction taken with public education efforts.

6.1.1 Two Basic Approaches - Piaget

Two basic approaches worth considering begin with a decision as to when and how to begin public education. If you subscribe to the belief that knowledge is best built on a "blank slate" or a fresh foundation free of fissures or foreign elements than you might adopt an approach that reflects the philosophy of the renowned developmental psychologist Jean Piaget and his theories of cognitive development. Put simply, Piaget believed that knowledge is the product of a series of sensory building blocks called schema. The schema contain bits of information that we gather through physical and emotional experiences.

To relate schema to recycling, most people in their late teens or older when handed an empty tin can and asked to identify it would respond that it is a piece of garbage. Logic would follow that in most cases, in the home, they would treat an empty can as though it were garbage and throw it in the trash. On the other hand, people who have been exposed to recycling from a very early age when asked the same question might respond that the can is something that can be recycled. In the home, they are likely to respond by tossing the empty container in the recycling bin or bag.

For communities interested in and willing to begin education with the very young, a Piaget oriented approach would work well. Teaching children, from the beginning of their educations (as early as age 2) about recycling and of ways that they can participate can set the stage for life long behavior. Obviously, beginning with very young children is a departure from many public education approaches, yet has the potential for long term reward. That reward would be realized as the community's youngest residents grow older, influence their families and friends and form the basis of a permanent behavior pattern.

6.1.2 Two Basic Approaches - Skinner

The second approach involves public education designed to change habits already in place. Some would look toward lessons taught by psychologist B.F. Skinner and his work in applied behavior analysis. Skinner's "science of behavior" maintains that all behavior is the result of a natural sequence of events. The events consist of an ANTECEDENT which yields a BEHAVIOR and then a CONSEQUENCE.

Antecedent-------Behavior-------Consequence
(prompt or cue) (reaction) (reward or punishment)

Related to recycling, the antecedent could be an article in a newspaper about mixed paper recycling. The article would prompt the realization that the reader has a choice of behavior when it comes to the disposal of mixed paper. A recycling behavior might be the accumulation of junk mail or other specified mixed paper and periodic delivery to a drop-off facility. The consequence could be a financial reward provided by the drop off center or simply an "atta boy" from the worker that runs the facility that makes the individual feel good about the visit. The "atta boy" is positive reinforcement which will be recalled the next time mixed paper becomes an issue in the home. The resident is likely to repeat the recycling behavior if the reinforcement was positive enough. Conversely, if upon visiting the recycling center the resident gets a flat tire due to debris on the site, the negative consequence that accompanied the visit will probably discourage this particular recycling behavior in the future.

If you plan to change the behavior of the people serviced by your program it is important to design the program around clearly defined recycling initiatives, intended residential behavior and positive consequences. A good example would be a program that seeks to recover type one plastic containers. Your public education effort must reach out to those that generate this type of material in their homes or businesses. Clear direction must be provided to enable residents to determine the difference between type one and type two containers, how you need them prepared and method through which they are stored and transported to the curb for collection. By clearly describing the activity and desired resident response, you enhance the odds of the behavior leading to the successful collection of the material. If there is a misunderstanding and plastic bottles are rejected at the curb, it will take but a few of these experiences to take the form of punishment rather than reward. If the resident feels that recycling efforts will result in punishment, the result could be a drop in participation.

6.2 Communication Do's and Don'ts

Research on motivating desirable waste management behavior conducted by Geller and Lehman and published in the Journal of Resource Management and Technology identified the following do's and don'ts. Regardless of education philosophy, these guidelines apply in most areas of motivational communication. By taking note of them, you will likely avoid heated debate with residents or elected officials in the future.

- Motivational signs, messages, or slogans are usually not sufficient to precipitate changes in behavior, unless the desired response is viewed as convenient and the cue occurs at the point of the desired action.

- General requests or recommendations may be inexpensive to field, but could be useless in changing specific behavior. Example, "let's all recycle" versus, "Please participate in our community recycling program by placing your newspapers, glass containers (clear, green and brown), number one and two plastic bottles and aluminum cans in the city provided, recycling container, at the curb the first and third Wednesdays of the month."

- Educational incentives are usually more effective when linked with audience participation via verbal exchange or hands on demonstration. A good example of an audience participation approach could be a material sorting game or a demonstration where children assist an instructor in making paper out of waste material. Lectures or video presentations that rely on passive listening are less effective.

- Strategies that set examples or provide role models for changes in behavior provide significant return on investment, especially when linked with support material echoing the principles of the example. One of the most common approaches could be called "practicing what you preach." Many communities adopt recycling within the administrative and operations branches of government. They adopt and promote polices which encourage recycling and waste prevention by government employees and establish guidelines for purchasing recycled or recyclable materials.

- Public commitments and goal setting approaches are very effective and, as communications strategies, inexpensive to use. Many school programs include the use of recycling pledge cards. These are used to solicit written commitments from students. One of the most common examples is the Recycling Ranger program used by many schools in New Jersey. After attending a recycling education program, children are made deputy Recycling Rangers (certificates or "membership" cards are provided).

- Strategies that focus considerable attention on the consequences of behavior, either reward or punishment, must be accompanied by clear directions as to the recycling initiative that prompts the behavior. If a promise of reward is made, yet not provided do to a misunderstanding of the recycling rules, program credibility may be lost or retarded until proper instruction is provided.

- Prompts with demanding instructions or unpleasant consequences should be avoided. The threat of negative reinforcement often yields behavior opposite of that desired. Recyclables rejected at the curb often result in hard feelings and resentment that are difficult to reverse.

- Individual perceptions of the type of reinforcement (positive or negative) may determine the impact on behavior and require careful consideration before committing to full implementation. If an elaborate positive reward is offered such as payment for material recycled, a long term commitment may be required, regardless of market conditions in order yield a long term change in waste disposal behavior. The sponsoring entity needs to carefully consider the true cost of such a program.

- Reinforcement linked to group behavior yield an additional level of social control and are desirable where feasible. Peer pressure is very effective when it comes to recycling. A good example is the impact of being the only house of the block to not put recyclables out for collection. There is usually a noticeable absence of participation the entire neighborhood can see.

- Perceived threats to freedom and negative psychological responses can be avoided by participatory planning approaches. Many programs are planned with the input of a public advisory group. Often referred to a Solid Waste Advisory Committees (SWAC), the SWAC is made up of individuals who reflect the businesses and residents of the community. The SWAC, media and periodic public hearings will help to avoid misconceptions or feelings of disattachment with the recycling process.

- Reinforcement should be provided on a long term basis and seriously enforced until residents adopt recycling behavior as a natural part of their routines. The goal is to assist the public in forming a recycling habit. That can be more difficult than is sounds because, generally, recycling habits will require a change in solid waste management habits that have been in place for the lifetimes of local residents.

- A specific entity should be clearly assigned responsibility and resources needed to provide reinforcement for recycling behavior over an extended period of time. Often referred to a the local recycling coordinator of solid waste director, the individual(s) assigned must be given clear goals and objectives as well as the resources needed to reach levels of expectation.

Experienced recycling coordinators could contribute additional do's and don'ts to this list. Probably to the most important thing that can be done is to design public education programs specifically to meet the needs and perception levels of the host community. Conversely, many communities have made costly commitments to recycling education and operational approaches on a "copy cat" basis. The belief that money can be saved by adopting the neighboring community's program is an invitation for disaster. Your neighbor may have copied its program from somewhere else with conditions entirely different from yours.

6.3 Where Do You Begin?

Before making plans or commitments to recycling education strategies it is important to determine the level of public awareness that exists within the community. Many recycling coordinators have assumed that residents either do not know anything about recycling or know and care as much as they do. Rarely will either scenario be true.

6.3.1 The Monmouth County Example

A good example can be found in Monmouth County, New Jersey. Monmouth has a population of nearly 600,000 and is located on the New Jersey shore about 30 miles from New York City. In 1986, work began on a mandatory residential and commercial recycling program that included all 57 communities within the county. Careful waste stream studies were conducted over a six year period beginning prior to the mandatory program and continuing over a period during which most municipal programs matured.

In the fall of 1992, decisions were made to reassess the progress of the recycling program. One of the tools used to conduct the assessment was a professionally designed and administered public opinion survey. The survey revealed a surprising lack of awareness of the problems facing the county relative to solid waste management and its residents' responsibilities relative to being a part of a solution. This response was ironic in light of an extensive, long term educational effort at municipal, county, state and national levels. Further, for nearly two years, the county was the site of heated debate relative to the construction of a material and energy recovery facility. Plans for the waste-to-energy plant were ultimately cancelled due to a loss of a referendum.

Upon review of the survey data, county officials changed course with their plans for a new wave of public education. Specific elements of the county population were targeted for additional attention and the overall structure of the recycling and waste management message was modified to fill critical gaps in knowledge and awareness. There was initial concern relative to the cost of the survey (approximately $20,000) and the time required to plan and conduct the program (four months). Of greater concern was the potential greater loss of funds and time had the program not been effectively targeted and the best message developed.

Many recycling coordinators will be faced with evaluating the results of past or ongoing public education efforts. While it may appear to be cheaper to guess or speculate as to effectiveness, in the long run it pays to begin with a properly designed and administered survey.

6.3.2 Public Opinion Survey

The Monmouth County Public Opinion Survey is a good example of a recent effort specifically designed to determine awareness, attitudes and opinions toward municipal solid waste issues including waste generation, reduction, recycling and disposal. The results of the study were used as the basis for the development of a communications plan for implementation by the County.

When designing a survey for any community, it is important to remember that recycling is a major element of a larger effort to manage municipal waste. Given the time and expense needed to implement a survey program, the tool should be used to assess perceptions and attitudes on a broader basis. For example, it would be tempting to gear the survey strictly to the measurement of public participation rates. By doing so, an opportunity would be lost to determine attitudes relative to waste reduction.

6.3.2.1 Steps to Planning a Survey - Objectives

As you plan your survey, it is helpful to first establish a clear set of objectives. Before they are finalized, share them with other individuals involved in recycling efforts and solid waste management in your jurisdiction. Make sure that key elected officials are aware of your plans and have an opportunity to contribute suggestions relative to the objectives.

The objectives that guided the Monmouth survey included determination of the following:

- The awareness level of existing solid waste management disposal problems and possible solutions;
- The level of knowledge of current solid waste management initiatives underway in the County;
- The current level of mandatory and voluntary resident recycling and/or composting efforts;
- Awareness and attitudes toward waste reduction options;
- The level of perceived public responsibility for waste generation and management activity;
- Attitudes toward recycling options including levels of participation and willingness to take on additional responsibility; and
- Where significant differences in opinions exist to allow for the development of a communication plan designed specifically to improve overall public participation and performance in waste management programs.

6.3.2.2 Steps to Planning a Survey - Methodology

Once the objectives have been identified, turn to the development of methodology for the survey. Care must be taken at this point not to mislead yourself as to the approach planned and the value of the subsequent results. Carefully consider retaining an experienced market research firm or other entity skilled in scientific polling and statistical analysis. Sample size will be critical to interpreting the reliability of the survey results. The market research firm can guide you through the development of the sampling approach and help you tie the results to a level of confidence and the margin of error based upon your sample size. Often random sampling is employed based upon telephone numbers from local directories. Generally, a basic inventory of numbers will be drawn, a portion of the first digits will be retained and random endings added. This approach takes into account unlisted and new numbers. Call-backs are administered during different times of day and evening to eliminate time-of-day bias. It is also helpful to build in quotas to the sample selection process. Type of housing unit, age, education, income and gender should also be considered. Care should be given to assure realistic splits for housing ownership versus renting, single family homes versus multi-family units surveyed, age across a number of age ranges, multiple categories for education, income on a broad range and gender.

6.3.2.3 Steps to Planning a Survey - Analysis

As the analysis is conducted, it is important to determine frequencies of responses for all question. Cross tabulations should be done according to demographics to determine significant differences in attitudes and to identify significant patterns.

After the data is analyzed, look for major differences in responses to the questions. The Monmouth County Survey analysis focused on differences relative to four key issues:

- Opinions on Garbage
  Who creates the most garbage in Monmouth County?
  Who has the most responsibility for solving the garbage problem?
  Should education on recycling and garbage disposal be included in the school curriculum?

- Awareness of Waste Disposal Problems
  Do you think Monmouth County has a problem disposing of its garbage?
  Would you suggest ways of reducing the amount of garbage thrown out in Monmouth County?
  What are some other good ways of reducing garbage?
  Are you aware of Monmouth County-sponsored mixed-paper recycling centers?
  Where do you get the most of your information about recycling?
  What materials are residents required to recycle in Monmouth County?

- Behavior with Regard to Recycling
  What materials (mandatory) do you currently recycle?
  What materials do you voluntarily recycle?
  Are you using the mixed-paper recycling centers?
  How are you Willing to participate?

- Additional materials on a voluntary basis?

6.3.2.4 Steps to Planning a Survey - Results

The conclusions of the survey will reveal the level of awareness within the population surveyed. It will provide evidence as to the need for public education and in what areas (demographic and programmatic). It will determine awareness relative to specific program elements as well as participation rates can be quantified. The survey can provide information that will serve as the foundation for waste reduction efforts. You will be able to provide a quantitative assessment of how well the existing programs/facilities are doing and identify where more work is needed.

Armed with survey results, you can target your communications plan to the actual needs that exist within the community. At the same time, you can use the findings to help in selecting the most cost effective media through which to deliver your message, the appropriate format and frequency. Your communication program will adopt a "rifled" approach versus a less effective "shotgun" marketing program.

6.4 The Communications Plan

The public education or communications plan is the road map to implementing your strategy. It should identify a target audience, the specific actions desired and desired responses. It will include a mission statement, a central theme, measurable goals and objectives, specific strategies including action items, schedules and budgets.

Media buys should be clearly identified including specific media form (i.e., print, radio, cable television), frequency and budgets. As part of the plan, provisions should be made to conduct periodic follow up surveys to gauge results and guide adjustments to the plan.

The budget should be broken into specific educational elements and the total budget conveyed on a cost per resident basis. Many programs are dependent upon free coverage in newspapers or through public service announcements on radio or television. Given the need for sustained educational information, dependence on free coverage may yield less than desired response. Funds should be set aside for paid advertisement if your program is focused on print or radio media strategies.

Large sums of money are not necessarily a requisite for effective public education. Some plans can be geared toward the custom adaptation of materials and strategies developed by others. Also, as mentioned under Do's and Don'ts, strategies based upon modeling or hands-on experiences are cost effective and can result in favorable results. As you develop a plan, pay close attention to the information your survey provides and avoid reinventing the public education wheel. There are lots of high quality materials available for a fraction of their original development costs.

6.5 Public Education Strategies

Your communications plan will be like a pie divided into slices or substrategies, each aimed at achieving the mission you identify in the beginning of your plan. Hundreds of substrategies have been developed by talented educators and recycling coordinators from around the country. Attend any conference on recycling or solid waste management and you will see new examples of classroom initiatives, the innovative use of videotape, mailings, poster contests, county fair displays, public service announcements, paid advertisements, recycling logos, characters and clever costumes. You will also learn about new or innovative waste reduction strategies, information sharing networks, trade journals, and free materials available via manufacturing organizations that represent paper, plastic, steel, ascetic and glass packaging interests. Free advice is also available through the National Recycling Coalition, the Solid Waste Association of North America and others.

In this section, three detailed examples of public information programs are offered. Each represents a comprehensive strategy and examples of substrategy. Each is either unusual or an example of a particularly effective approach. Space limitations restrict the number of detailed examples offered, however, it would be safe to assume that there could be as many as there are creative recycling coordinators in our country.

6.5.1 The Clean Community System

Keep America Beautiful, Inc., a nonprofit organization originally founded to address litter prevention and community involvement in solid waste issues, developed the Clean Community System (CCS) in the mid-1970's. The system has proven to be one of the most effective approaches to changing human behavior on a community wide basis. KAB has based its model on the promotion of practical collaboration between local government, business and civic organizations aimed at solving solid waste management problems. The Clean Community System has been implemented in more than 400 U.S. communities in nearly every state. In addition, it has influenced litter-control programming in six other countries: Australia, Bermuda, Canada, Great Britain, New Zealand and South Africa. The CCS model provides one of the best working models for organizing, motivating, and evaluating community action toward the large scale solution of environmental problems.

Motivational strategies particularly the use of heavily promoted awards programs, are important to the CCS approach. Positive reinforcement via social approval and public recognition are the primary rewarding consequences. The achievement of "Certified Clean Community" status are major potential motivators for mayors, municipal staff, and volunteers. Figure 6-1 illustrates the organizational framework for a comprehensive solid waste management program adapted from the KAB plan for the Energy Conserving Community System.

The Resource Community System (RCC) proposal is a two-stage plan through which communities initiate waste management procedures including recycling and waste reduction in order to earn eligibility as a "Certified Resource Conserving Community" (RCC), and then specific criteria are developed for qualification as a "Responsible Resource Community" (RRC).

As under the KAB plans, completion of each stage would result in heavily publicized praise and if possible, financial rewards through government and industry grants. Positive reinforcement must be applied at every stage of the process in order to encourage progress to ultimate recognition and the achievement of the mission. Additional incentives will be needed in order to maintain performance levels after the community has achieved RRC status. From a political perspective, elected officials may strive to be certifiably responsible with local resources. The potential for this model extends far beyond recycling to responsible government and conservation.

As illustrated in Figure 6-1, the process begins with a letter of endorsement from a mayor, county executive or other ranking elected official. With that letter comes a commitment to send a work team to an RCC training conference and to provide sufficient funds for the first year of operation. Training conferences can be planned and executed via a county, state, or university extension system. Under the KAB model, local training entities are accountable to a national coordinating organization. The work team representing the community is made up of individuals from government, commercial and community organizations and related volunteer groups.

At the training sessions, the community work teams learn about methods for coordinating, motivating and evaluating a local program for recycling and/or solid waste management. The team assembles its own plan designed to address the needs of the residential/consumer, governmental/institutional, and commercial/industrial sectors. During workshop training, attendees learn about waste prevention, reduction, recycling and proper waste disposal. The KAB workshops focused a significant amount of attention on litter control. During the same workshops, attendees are asked to work with RCC team members to develop an incentive/reward program for prompting and encouraging long term efforts in the desired behavior.

An Organizational Framework for Certification as a Resource Conserving Community and Resource Responsible Community

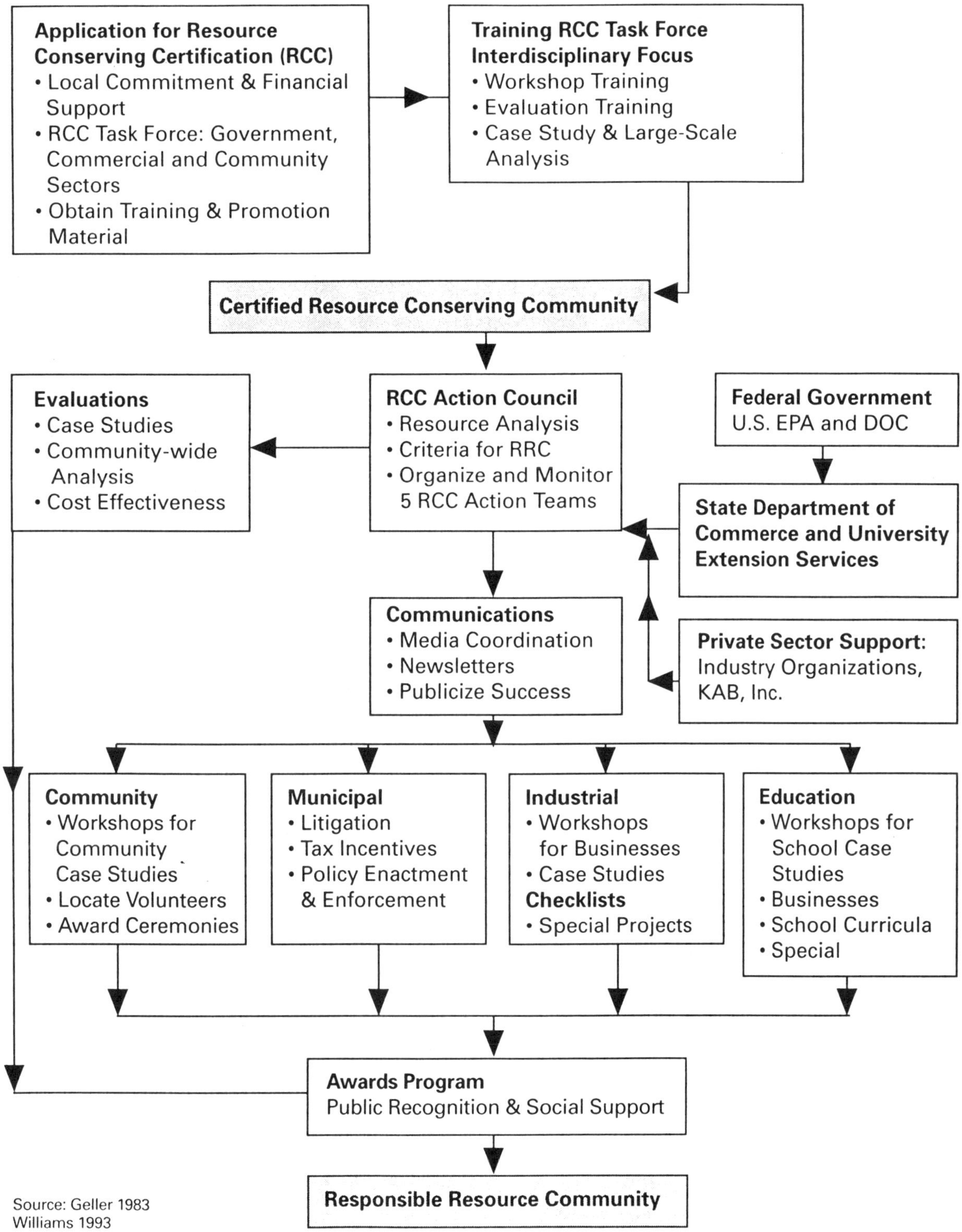

Source: Geller 1983
Williams 1993

After a week of training in the workshops, the work teams are asked to implement a large scale program in their home regions. Communities that participate in the workshop training are certified as "Resource Conserving Communities". Progress by the local work team is monitored and rewarded by local, state and federal entities. The ultimate goal is that of recognition as a "Responsible Resource Community" (RRC). Prior to achieving RRC status, the RCC work team must establish a local headquarters; receive funding commitments; present RCC workshops to key local leaders in government, institutions, commercial activity, industry, and community agencies. The RCC must select demonstration projects representing local needs and organize an RCC Action Council.

Under the KAB model, the Action Council is essential from both an organizational point of view and as a source of workers for implementation of the RCC plan. Typically, members of the council would come from the residential/consumer, government/institutional and commercial/industrial sectors. In addition, individuals with backgrounds in litter control, recycling, economics, education, engineering, psychology and sociology would also be recruited. Priorities for the council would include the establishment of a resource center for information relative to recycling and solid waste management; provision of a communications link to local experts who administer the RCC training programs, offering certification as RCCs and RRCs; organizing and monitoring the RCC action teams responsible for waste management education, promotion, and demonstration projects; evaluating progress of the RCC action teams; and developing criteria for a responsible resource community (RRC).

KAB suggests that the Action Teams be organized as follows:

- The Communications Team would be responsible for publicizing the activities of the RCC, providing for positive reinforcement of individual and group activities efforts to encourage resource conservation and other efforts that increase recycling. All communications media should be used and an internal newsletter developed.

- The Education Team would be responsible for keeping all other teams informed of emerging issues or techniques for solid waste management and for promoting recycling in public schools. It may also sponsor public education programs for teaching waste management principles including waste reduction, recycling and reuse techniques.

- The Community Team would be responsible for locating and coordinating local civic organizations in resource conservation and recycling initiatives. This committee would train civic volunteers to teach workshops and initiate neighborhood projects and assist in evaluating community programs. It would also conduct award programs to provide additional positive reinforcement for individuals and groups working toward RCC goals.

- The Government Team is responsible for examining government policies and encouraging those that promote sound planning and resource conservation efforts over a long-term basis. This team could suggest policy, programs and legal initiatives aimed at motivating waste reduction, recycling and responsible disposal practices. The Government Team would also plan and execute government workshops on waste management issues for public employees.

- The Industrial Team is charged with promoting waste reduction, recycling and sound disposal practices in the commercial/industrial sector of the community. It should establish prominent examples of practical and appropriate waste management practices via the private sector. It should develop and sponsor workshops on waste reduction, recycling and waste management for groups in the commercial/industrial sector.

While the KAB model provides a comprehensive framework on which hundreds of successful programs have been structured, the most important factor to the success of this model has been the coordinated approach through which it is administered. Each waste generation sector is recognized, provisions are made for training, follow up measurement and significant emphasis placed upon positive reinforcement. Roles are incorporated for multiple levels of government and the use of existing mechanisms like cooperative extension services and state universities to assist in educational efforts. KAB was successful in implementing its model on a national scale. Many recycling coordinators have access to local KAB affiliates and their trained professional staff, as well as, extensive education resources available at little cost.

6.5.2 Beginning with Preschoolers - The MISTER ROGERS Model

There are at least two major factors that can influence a community's ability to reach higher recycling goals. The first is the community's ability to align programs with recycling markets and that those markets will be able to absorb recovered material over an extended period of time. The second factor is public attitude and willingness to participate in the solid waste management program on a long-term basis. In many cases, public attitude must be compatible with dramatic increases projected in recycling and reduction efforts over the long term, 20 year planning period.

This example, which reflects Jean Piaget's theory discussed earlier, was designed to take advantage of a "clean slate" approach to education. By focusing on very young children, ages two to four, a foundation is constructed for behavior responses that can last a lifetime.

Preschoolers between the ages of two and four are ready to begin learning about recycling and waste management. In 1989, public television's MISTER ROGERS' NEIGHBORHOOD featured a five-part series of programs on the environment. One of the programs (No. 1617) was specifically tailored to convey a message on recycling and conservation. The 1989 effort was the first major initiative aimed at carrying a recycling "lesson" to very young children. Aired on nearly three hundred stations of the Public Broadcast Service, more than 14 million households tuned in to view the program. Since its debut on Earthday of 1989, the series of programs has been broadcast nationwide ten times.

Briefly, the program opens with Rogers walking into his house carrying a bag filled with glass bottles and cans. He talks about recycling and sings, "I like to take care of you." His friend stops by and helps him separate recyclables into two boxes. Together, they visit a recycling center where they talk with the workers and watch what happens to the bottles , cans and other material. Upon returning home, the program transitions into a puppet segment where the discussion shifts to a local garbage problem. The characters begin to search for a new landfill and some suggest that recycling would help the situation.

In the final program segment, Mister Rogers asks his viewers to think about what we really need before we buy and think about other things they can do with waste material before we throw them away. This final request has major significance. Preschool children love and respect Fred Rogers. A poll by Playskool asked preschoolers in five American cities what famous person should be the next president, 45 percent said Mister Rogers. If you share Fred Rogers' opinion that "attitudes are caught, not taught" you might conclude that the program described above provides a broad reaching opportunity for millions of Americans to "catch" a recycling and waste reduction attitude.

Research by several reseachers would support a theory that the widespread application of the recycling and conservation message contained in the MISTER ROGERS segment could have a major impact on local behavior. The researchers used 10 to 15 minute segments from MISTER ROGERS' NEIGHBORHOOD to test groups of small children. They found that after viewing the clips, the children were found to have improved in task persistence, self control, and tolerance of delay in comparison to children who viewed either anti-social or neutral films.

The use of the program as part of a nationwide, comprehensive public education strategy could yield significant, long lasting results. Our research concluded that if a national effort were launched in 1990, by the year 2010 over 76 million Americans, or 27 percent of the nation's population could have been exposed to a positive message relative to recycling and conservation. Even under the most ideal circumstances, the program could not reach every child directly. Language, location, economic resources, and parental attitudes pose specific obstacles. The residual effect of the program could help to offset the obstacles.

Public attitude is almost certain to be influenced by such an initiative. Within this century, society has shifted its approach and attitude relative to waste disposal. Not long ago, a parent would ask a child to take out the garbage. The son or daughter complied by carrying the waste to the front room window and dumping it onto the ground below. By the year 2010, assuming significant strides are made in public education, when a father makes the same request, we suspect that the child will automatically spend more time and effort preparing the waste to enter the management cycle.

How do you take this example and apply it on a local basis? If used to the greatest advantage, the MISTER ROGERS' material including the videotape and an accompanying 24 page activity guide, could provide the comprehensive strategy needed for long-term results. The substrategies could include a publicity program entitled, "Starting at the Beginning", focusing attention on the need to begin good waste management and conservation habits early on. Publicity could promote frequent local broadcasts on public access television (with specific permission from Family Communications Inc.). It could promote the use of the video and work books in local preschools, kindergarten and first grades as well as library story hours. The guides could be distributed to homes with young children. Poster competitions can be designed around the "Starting at the Beginning " theme as could advertising campaigns.

Each of the above mentioned substrategies have been used by communities across the country. Aside from the obvious educational benefits, sponsoring communities benefited from the high level of name recognition that accompanies the material and by the fact that the material is available for any noncommercial use to the general public. Users are asked to pay fees that roughly cover the cost of reproduction. All developmental costs were underwritten via philanthropic efforts of HDR Engineering, Inc. and the Sears Roebuck Foundation.

6.5.3 Mt. Pleasant User Fees

Another good example of public education efforts aimed at waste reduction, recycling and waste management practices can be found in Mt. Pleasant, Michigan. The community has a population of about 25,000, a large portion of whom are students at Central Michigan University. Like many communities, Mt. Pleasant paid for waste management activities through its operating millage. As the cost of waste management increased and out-of-county disposal became a reality, the city decided to switch from a tax based approach to a user fee system. The system is in many ways simple. Before user fees, there was a six bag per week limit on waste collected. Now there is no limit and residents pay for the exact amount of service they use. By the end of 1989, ten refuse bags and tags could be purchased for $6.00. Yard waste bags were available at ten for $2.50.
According to most researchers, public education was well served because the objectives of the user fee approach were:

- To make residents aware of solid waste management costs; and
- To encourage residents to reduce waste volumes and increase participation in recycling programs.

The approach worked. Isabella County opened a recycling facility in August of 1987. From that date until May, 1989, the facility diverted approximately two to three percent of the county's waste stream. Since Mt. Pleasant implemented its user fee system in June of 1989, participation at the facility increased by 50 percent and the amount of waste placed out in Mt. Pleasant for pick-up has dropped off dramatically.

Mt. Pleasant is an example of a program based largely upon the ABC Model discussed earlier. In this case, the city called for a change in behavior that was in essence to reduce the amount of waste placed out for collection and disposal. In order to comply with the request, residents had two choices that would yield positive reinforcement. The first was to adopt waste reduction strategies. The second was to make use of the local recycling facilities. Those residents that made either or both choices were rewarded with less garbage to haul to the curb, less cost for disposal and the avoidance of a ticket for noncompliance. Those that ignored the city's requests were the recipients of punitive reinforcement in the form of economic penalties and the task of buying bags and tags on a more frequent basis.

The educational campaign started several weeks before the new user fee program. City officials and local solid waste advisory committee members joined forces to launch a major educational campaign to explain the law.

Volunteers made presentations to local service organizations and community groups. Explanations of the new law were provided as well as hands on demonstrations on how to prepare recyclables. Press releases and public service announcements were used to reach television, radio and print media. Representatives appeared on talk shows. Paid advertisements were placed in local media. The city published two information brochures including how to's and alternative ways to dispose of solid waste. The city also produced slide presentations and flyers focused on composting, household hazardous waste and consumer buying habits. Six city sponsored open houses were held for local residents to learn more about the ordinance and related issues. The last substrategy focused on the distribution of a sample garbage bag, one free tag and program information materials to all single family houses.

The three examples shown above provide a broad range of strategies and substrategies. While most recycling professionals are hampered by financial constraints, they are generally creative individuals who find new ways to conquer their "mountains" each day. One of the attributes most critical to recycling coordinators is the ability to locate and borrow materials or resources developed elsewhere.

6.6 Varieties of Educational Methods

One of the most popular outlets for recycling education is the public school. In 1991, more than 148,000 California students from the state's 15 most heavily populated counties were the focus of a two week recycling education program. According to Roger O. Scott, head of the Southwest Regional Laboratory (SWRL) which developed the recycling education program, almost 5,000 teachers and 232 public and private schools used the curriculum materials. About 60 percent of the participating teachers taught kindergarten through third grade. The program included a series of ten 25 minute lessons emphasizing four objectives: the reasons recycling is important; the kinds of containers that can be recycled and redeemed; collection procedures for containers; and locations for recycling. Students were given color activity booklets and they listened to a tape and viewed a film strip explaining recycling. Beyond the structured program, teachers and students at 95 percent of the participating schools collected recyclables as an opportunity to apply what they learned.

In an attempt to measure the results of the initiative, SWRL conducted a survey with about a 75 percent response rate. SWRL reported that more than 94 percent of teachers indicated that their students enjoyed learning about recycling and more than 92 percent said that their students became more aware of recycling. Parents surveyed agreed that their children enjoyed learning about recycling. According to Scott, 80 percent of the parents surveyed reported that their child recycled more and 70 percent said that their families are now recycling. There is little doubt that children carry the recycling message home.

While the California results were impressive, recycling coordinators should keep in mind that getting teachers interested in recycling education can be a problem. Teachers are busy and often overwhelmed by the number of requests and mandates they receive to include specific instruction in an already crowded school day. Some teachers may not have the time within or outside of the classroom to research new topics and gather activity material. Tight school budgets present an additional constraint that may even preclude the cost of material reproduction.

As you plan to gain access to local students through the local school system, make sure that you provide as complete a package as possible to teachers. It should be age appropriate, of a professional quality and easy to read. For items intended for distribution, a sufficient number of copies should be available. They should be sensitive to a need for ethnic balance and other socio-economic concerns. Finally, do not forget that your material will have to fit into a specific time slot, probably no longer than one-half hour.

The examples mentioned earlier in this section have provided a broad range of applied approaches. It will be helpful to consider the following general list of substrategies as you prepare community specific programs. A survey conducted by the Environmental Hazards Management Institute (EHMI) in Durham, New Hampshire identified 30 educational methods, broken down into four general categories. While some may be familiar, there are sure to be a few that prompt new strategies for recycling coordinators:

- Personal Contact
  - Contact with recycling program staff
  - School workshops
  - Community workshops
  - Volunteer network
  - Community events
  - Block leader programs
  - Canvassing

- Print Media
  - Newspaper announcements
  - Newspaper advertisements
  - Newspaper articles
  - Bumper stickers
  - Recycling center signs
  - Recycling truck signs
  - Other signs
  - Newsletters
  - Mass-mailing of brochures
  - Publicly available brochures
  - Mass-mailing flyers
  - Publicly posted flyers

		Recycling bin stickers
•	Electronic Media	Television advertisements Television interviews Radio advertisements Radio interviews Telephone hotlines
•	Other	Recycling surveys Recycling committees Recycling awards Waste reduction curriculum Other (as specified by community)

One important and potentially fruitful area not revealed by the EHMI survey has to do with the use of public access television and the availability of production facilities and air time via most local cable television stations. For example, the City of Jacksonville, Florida capitalized on local resources to address a complex public education problem. The city had experienced difficulty in providing credible public education in the inner city. Residents felt disconnected with recycling initiatives and found it difficult to relate to a suburban model for recycling. A gifted recycling coordinator developed a concept for a recycling video to be written and produced in the inner city by inner city residents. The product was a very effective videotape featuring residents and an artistic flavor that reflected inner city culture. The production was largely made possible by access to cable television facilities. After the video was complete, public access stations provided a free and frequent link to residents of the entire city.

6.7 Other Sources of Public Education Information and Materials

Since the middle 1980's hundreds, if not thousands of recycling education programs and materials have been prepared. Much of the material is available to the public for free or at nominal charge. Most state departments of education have collected material or even developed curriculum specific to the population within their particular borders. There are also many state oriented recycling associations that assist local recycling coordinators with program development. Groups like the National Soft Drink Association publish extensive resource lists which often include specific material, directions for acquisition and price lists (202)463-6700. The National Recycling Coalition (202)625-6406 and the USEPA Public Information Center in Washington, D.C. (202)382-2080, as well as the Solid Waste Association of North America in Silver Spring, Maryland (301)585-2898 are also excellent sources of information. Local waste haulers and material recovery facility operators often have high quality educational materials available and may even be willing to sponsor or include recycling coordinators on local educational efforts. There is generally little need to "reinvent the recycling education wheel", however, great care must be taken to insure that, through efforts to save money, misinformation or inappropriate information is not disseminated to your target audience. Keep in mind that quantity may not be the same as quality when it comes to designing an effective, long lasting recycling education program for your community.

-7-

RECYCLING ECONOMICS

By implementing recycling programs, communities often face difficulties in financing the introduction of new services requiring both capital investments and increased operating expenses. These new services may result in increased tipping fees at solid waste facilities, increases in property assessments, funding through existing solid waste system programs, funding through different types of bond financing or private loan placement arrangements, and requiring state grant funds to prevent any adverse impact on local community budgets. This section focuses on a general overview of the costs currently anticipated for recycling programs and discusses the major sources of funding for these programs.

7.1 Costs of Recycling Programs

Over the last number of years, numerous communities have implemented various kinds of recycling programs ranging from drop-off centers to communitywide curbside collection systems. However, it is often very difficult for decision-makers in other communities to compare the costs of their programs with those of another community because of differences in the way costs are accounted for in each local program. Very often, data are not identified on who pays, over what time period costs are incurred by the community and how much volunteer labor enters into the overall costs reported by each community. Also, data are generally limited on the portion of a community's overall solid waste system costs that are avoided by recycling. Lastly, data presented by different communities often do not indicate the market prices received for selected recyclables. As shown in Table 7-1, prices for recyclables do vary dramatically by region.

Contrary to popular expectations, recycling programs savings from avoided disposal fees and revenues from the sale of recyclables often do not offset the costs of implementing and operating a full-scale, community recycling program, particularly when local market conditions for recyclables are depressed. Under these conditions, recycling programs must be supported by a variety of local or state revenue sources. For example, a study undertaken by the Commonwealth of Massachusetts in 1992 estimated that up to $30 million per year would be necessary in the short term to assist local governments in Massachusetts in developing mandatory curbside collection programs.

In contrast, a study undertaken in 1993 by the Clean Washington Center, a state agency responsible for recycling market development in the State of Washington, reported that recycling was less expensive than disposal costs in four Washington cities (Bellingham, Seattle, Spokane, and Vancouver) which were studied, considering collection, transfer, and processing costs as well as the revenues received for the recyclables. Disposal costs in these four cities exceeded recycling costs and these cities have experimented with the use of innovative contracts with their private solid waste collectors to encourage route efficiency and volume-based fees to homeowners to reduce waste volumes by encouraging recycling. Again, it must be emphasized that the conclusions reached in this study may have limited applicability outside of the Pacific Northwest because of their proximity to excellent recyclables markets in the Pacific Basin which currently do not exist elsewhere.

7.1.1 Curbside Recycling Programs

Over the past few years, it has been reported that more than 5,000 recycling programs have been implemented in the United States, with many of these being curbside collection programs. Curbside recycling offers the greatest convenience and opportunity for materials recovery, but it is a labor intensive form of recycling. Covering costs of a curbside program with revenues from material sales is difficult to achieve. Therefore, evaluating cost-effectiveness of curbside programs should take into consideration costs for refuse collection, transportation and disposal, and environmental benefits such as landfill space savings.

Planning and estimating the cost of providing curbside collection is a major component of developing a curbside recycling program. Vehicle selection, fleet sizing, staffing and routing are the primary elements to be determined. In the past, recycling collection vehicles have been a myriad of vehicles from old school buses to flat bed trucks. However, in recent years, recycling collection vehicles have become more sophisticated. Compartmentalized collection vehicles, designed with consideration of loading height, storage capacity, movable compartment dividers, stand-up driving, and safety features, among others, are quickly becoming the norm for curbside programs. Although their initial capital costs are more than makeshift vehicles, greater vehicle productivity, collection efficiency, and reduced operation and maintenance costs are the overall benefits.

In general terms, fleet sizing will be determined through analysis of vehicle productivity, truck capacity, and collection efficiency. Vehicle productivity is defined as the number of stops passed in one productive collection hour. A "stop" is an eligible housing structure. Not all eligible housing structures will put recyclables out each collection day. Housing density, curb miles, traffic patterns, topography, climate and daily weather conditions, and degree of sorting required by the operator will affect the productivity.

TABLE 7-1

END MARKET REGIONAL PRICES FOR SELECTED RECYCLABLES

Commodities	Price/Ton	Commodities	Price/Ton
Old Newspaper:		Old Corrugated:	
Northeast:	$10-25	Northeast	$15-25
East Central:	0-15	East Central:	20-25
South:	0-25	South:	35-40
West:	12-20	West:	40-50
Clear (flint) Glass:		Amber (brown)	
Northeast:	$26-30	Glass:	$15-26
East Central:	40-50	Northeast:	22-25
South:	50	East Central:	50
West:	40	South:	20
		West:	N/D
Green Glass:		Steel Cans:	
Northeast:	$0-15	Northeast:	$65
East Central:	0-15	East Central:	30-60
South:	10-20	South:	40-60
West:	10	West:	40-60
PET:		HDPE Natural:	
Northeast:	$120-140	Northeast:	$100-200
East Central:	120-200	East Central:	80-120
South:	80-100	South:	120-160
West:	120-160	West:	0-100
HDPE Mixed Color:		Aluminum Cans:	
Northeast:	$80	Northeast:	$780-800
East Central:	120-140	East Central:	700-780
South:	80	South:	780-800
West:	60	West:	780-840

Note: All prices are per ton, end-market prices; N/D - no data.

Source: Recycling Times, November, 1993.

Truck capacity will be helpful in determining route size, and when a vehicle will have to leave a route because it has reached capacity. An estimate of average quantity of collected material in both volume and weight per stop per collection period will be needed to calculate how many stops a truck can pass before filling. the truck capacity along with an estimate of non-productive time (travel time to processing facility, unloading and break time), productivity rate, and the total number of houses served can be used to determine how many collection vehicles will be needed. Finally, the cost of labor to sort the recyclables at curbside and drive the collection vehicle must be factored into the overall collection cost for the recyclables.

Although many surveys of curbside recyclables collection have been conducted in recent years, there have been widely divergent claims as to the cost effectiveness of curbside collection programs. Unfortunately, much of the data reported in the trade press remains undocumented and are of limited usefulness because not all direct and indirect costs are generally reported so that an "apples to apples" comparison can be made. Generally, however, curbside collection programs nationwide have typically ranged from $0.50 to $2.50 per household per month. This cost is based on a community-wide program and all the households in the community receiving curbside service. The cost is generally distributed over all the households receiving service, regardless of whether the individual home participates in the program. The cost variance in these curbside programs are due to differences in the level of service provided, efficiency of collection crews, and quantities of materials collected. Table 7-2 presents estimated curbside collection costs for communities with populations of 30,000 and 200,000.

7.1.2 Drop-Off Centers

The cost of establishing and operating permanent drop-off centers is dependent on the level of processing, the amount of assistance provided, and the types of materials recycled, as well as the kind of collection and transportation furnished. Collection and transportation are usually the major cost factors for drop-off centers. In general, the operation of drop-off centers is economically favorable. Collection of materials can be handled in a number of ways: route planning and storage capacity at the drop-off centers may allow a routine pick-up in combination with a direct trip to the buyer; another typical method is to make pick-ups from each site and store material at a central location until a sufficient amount is gathered for a haul to the buyer. Using free land, volunteer labor, and unattended centers can substantially reduce annual operational costs.

TABLE 7-2

CURBSIDE COLLECTION COST COMPARISON BUDGETARY COST ESTIMATE

Item	Cost ($)	
Community Population	30,000	200,000
Households Served[1]	8,500	56,400
Capital Costs		
Vehicles[2]	$165,000	$ 907,500
Containers[3]	46,500	310,200
Total Capital Cost	$211,500	$1,217,700
Debt Service[4]	41,300	238,000
Operating Costs		
Labor		
Collection Crews[5]	140,000	770,000
Administration	35,000	70,000
Administrative Support	8,600	25,000
Maintenance	8,800	35,000
Total Labor	192,400	900,000
Equipment Maintenance[6]	21,100	121,800
Education/Public Relations	8,500	56,400
Subtotal	222,000	1,078,200
Contingency (20%)	44,400	215,600
Total Operating Cost[7]	$266,400	$1,293,800
Total Operating Cost + Debt Service	307,700	1,531,800
Collection Cost/Household/Month	$3.01	$2.26

Notes:
(1) Rounded to the nearest 100.
(2) Two vehicles estimated for city of 30,000; 11 for city of 200,000.
(3) One container per householder plus 10% for replacements.
(4) Assumes seven-year amortization at 8.5%.
(5) Two persons/truck.
(6) Equals 10% of equipment costs.
(7) Totals may not foot due to rounding.

Source: HDR Engineering, Inc., 1991.

TABLE 7-3

DROP-OFF PROGRAM COSTS
BUDGETARY COST ESTIMATE

Item	Cost ($)
Capital Costs	
Containers[1]	$ 40,000
Rolling Stock	65,000
Total Capital Cost	$105,000
Debt Service[2]	20,5000
Operating Costs	
Personnel	20,000
Maintenance	10,500
Program Promotion	10,000
Total Operating Cost	$40,500
Total Operating Cost + Debt Service	61,000
Revenues, 500 tons @ $27/ton[3]	13,500
Net Costs[4]	$ 47,500

Notes:
(1) Assumes four sites, each with six igloo containers, and one large, enclosed container for newsprint.
(2) Amortization of containers and rolling stock in seven years at 8.5%.
(3) Assumes recovery of 70% newspaper, 27% glass and 3% aluminum at currently available per ton market prices--$5 for newsprint, $20 for glass, $600 for aluminum, $100 for plastics, $50 for ferrous.
(4) Total operating costs plus debt service less revenues.

The costs of a drop-off program can vary significantly because of the wide variety of system designs. These designs can range from highly elaborate centers which also serve as environmental education centers to simple sites consisting of a few containers. Table 7-3 presents cost estimates for a four-site drop-off program that includes separate containers for aluminum, plastics, glass, and ferrous, as well as a roll-off container for newsprint.

7.1.3 Materials Recovery Facilities

The implementation of curbside recycling collection programs has resulted in a dramatic growth in the construction of facilities to sort and process collected recyclables for their end markets. These processing facilities, called material recovery facilities (MRFs), improve the marketability of the collected materials for sale to end users, as well as allowing the community to accumulate the recyclables into larger lots which enhances the overall price received from the various end use markets. In order to accomplish these goals, MRFs incorporate a variety of processing equipment which varies in complexity. For example, some MRFs utilize a small tipping floor where materials are manually separated. Other larger facilities incorporate the most modern automated equipment utilizing trommels, screens, magnets, and similar sorting systems. However, whether the system is simple or high tech, all MRFs utilize some form of manual labor and generate several major cost items. For a processing center, major cost items are buildings, processing equipment, rolling stock, and safety and environmental protection equipment. Operating costs include labor, equipment maintenance, administration, and material marketing.

Since the growth of MRFs has only been a recent development, there are limited data available on capital and operations costs which can accurately be compared from one location to another. In 1993, the National Solid Waste Management Association (NSWMA) commissioned a study to focus on this apparent need for detailed processing cost data from operating MRFs. Target candidate facilities throughout the United States were contacted and sent a detailed questionnaire which requested specific cost information. This information was analyzed and ten facilities which supplied all of the requested information were selected for further field research. These facilities represented the geographical pattern of existing MRFs; four of the MRfs were located in the Northeast, three were in Florida, one in the Midwest, and two were located on the West Coast. These facilities also represented the variety of automated and manual processing methods and equipment.

The study reported the following important findings:

- The average cost to process recyclables at a MRF, before revenues from the sale of recyclables are considered, is \$50.30 per ton, with a range of costs from $28.11 to $72.06 per ton. The median cost is $48.54 per ton;

- Paper costs $33.55 per ton on average to process, considerably less than commingled recyclables, which cost $83.36 per ton on average;

- Commingled bottles and cans have a much greater range of processing costs, from $40.76 to $146.29 per ton than paper, which ranges from $20.43 to $55.93 per ton;

- Newspaper is the least expensive recyclable to process, averaging $33.59 per ton. Mixed paper, $36.76 per ton, and corrugated boxes, $42.99 per ton, also have relatively low processing costs (see Table 7-4);

- Plastic is the most expensive of the typical recyclables to process, averaging $183.84 per ton for PET and $187.95 for HDPE (see Table 7-4);

- The costs of processing clear and mixed glass are generally lower than the costs of processing amber and green glass. For most MRFs, this is because clear and mixed glass are more abundant in the waste stream and typically are being removed through negative sorting or mechanical means;

- Aluminum cans have the greatest range of costs, from $72.88 to $362.59 per ton. Newspaper has the narrowest range of costs, from $19.94 to $55.33 per ton (see Table 7-4);

- Worker productivity is greatest for paper, averaging 7.42 tons per employee per day. Metals account for the second highest productivity, at 5.96 tons per employee per day followed by glass, 4.21 tons, and plastics, 1.57 tons;

- Labor is the highest cost component, accounting for over one-third of the overall processing costs on average; and

- At current market prices offered by end users for most recycled materials, average costs exceed revenues. While the average processing cost is $50.30 per ton, the average value of a ton of recyclables is approximately $30.

TABLE 7-4
RECYCLING PROCESSING COSTS

Material	Range of Costs ($)	Average Costs ($)
Newspaper	19.94 - 55.33	33.59
Corrugated (OCC)	20.29 - 56.26	42.99
Mixed Paper	16.82 - 65.59	36.76
Aluminum Cans	72.88 - 362.59	143.41
Steel Cans	30.22 - 125.64	67.53
Clear Glass	37.17 - 105.62	72.76
Amber Glass	69.70 - 148.92	111.52
Green Glass	57.56 - 134.21	87.38
Mixed-Color Glass	28.51 - 76.24	50.02
PET Plastic	64.43 - 295.35	183.84
HDPE Plastic	121.58 - 256.15	187.95

SOURCE: NSWMA, 1993

However, the data presented by the NSWMA study only tells part of the story about the economics concerning the operation of MRFs. While for many facilities costs of processing recyclables currently exceed revenues received, many communities and operators take into consideration the avoided tipping fees at alternative disposal facilities to conclude whether such facilities are profitable. The data do show, however, that MRF operators can greatly enhance the potential profitability of their facilities by altering the mix of materials processed and eliminating materials for which a market does not currently exist.

7.1.4 Yard Waste Recycling Programs

In general, accurate and reliable cost data on yard waste collection programs are difficult to obtain. Table 7-5 illustrates the range of estimated costs and revenues for different yard waste recycling programs across the United States. The costs and effectiveness of a yard waste recycling program depend on many factors, including yard waste separation, collection and processing techniques, material storage and marketing requirements, administrative costs, and participation by residents.

Further, due to seasonal variations in yard waste generation and inconsistent participation by some households, it is often difficult to accurately predict actual participation, diversion rates, and overall costs based on data collected in other communities. It is for this very reason that pilot yard waste collection and processing programs are necessary in communities considering the implementation of such programs to accurately predict the ultimate costs of the overall communitywide program. For example, many programs have found that debagging garbage cans or plastic bags at curbside, which was initially selected at the outset of their program, proved inefficient and cumbersome in actual practice. Consequently, they were often forced to bring yard wastes into their yard waste processing facility bagged and accomplish debagging as part of their normal processing procedures. However, while labor costs increased somewhat because of manual or automated debagging, they found that the efficiency gained in keeping collection costs steady during the peak season offset these additional costs.

The costs for composting can be typically grouped into capital costs and operation and maintenance costs. Capital costs include such items as land, construction, buildings and equipment and project implementation. These are usually amortized or depreciated over the assumed life of the project. Operations and maintenance includes ongoing costs for labor, fuel, utilities, materials, supplies, overhead, and compliance with environmental regulations. These costs are often shared with other public agencies which makes it sometimes difficult to compare the overall costs of one location with that of another.

One of the most comprehensive studies on the costs of yard waste compost was performed by Pinellas County, Florida. The county undertook this evaluation to determine its incremental costs for expansion of its yard waste recycling program. The objective was to determine what its production costs would be if subsidies such as loaned labor and grant funds were eliminated and the County contracted with a private vendor for operation of their project. Participating cities in Pinellas County paid the county a tipping fee of $15 per ton with state recycling grant monies contributing an additional $7 per ton toward labor and equipment for windrow processing, reject disposal and mulch distribution to cover the overall cost to the county of $22 per ton. As shown in Table 7-6, the key elements in this budget estimate include county administration, labor, operation and maintenance, replacement of equipment, contractor equipment rental, contractor disposal of rejects, contractor profits, and mulch distribution. the $29.94 cost per ton was based on a production rate of 25,000 tons per year.

Benefits derived from yard waste recycling include potential revenues from the sale of finished product, the avoided cost of landfilling or incineration, and the avoided cost of using the mulch or compost as a substitute for some other product on public works projects instead of buying other material for the same purpose. A recent EPA study targeted eight yard waste composting projects sponsored by communities across the United States. One program sold the product for as much as $25 per ton to local landscapers and nurseries. However, the report does not specify what quantities were sold at this price. Most often, yard waste composting projects realize minimal to zero revenues, claiming benefit from the avoided cost of landfilling if the product is taken away by an end-user or from the avoided cost of topsoil if the product is needed as landfill cover.

7.2 Financing Programs

Once the alternatives for development of a recycling program have been evaluated, debated, and decided by its decision-makers, each community must then find the way or ways to finance the implementation of the favored program. There are a variety of programs used by local government for funding of the initial capital costs of recycling facilities and overall programs, as well as their operational costs. These can range from funding the program through general revenues of the community or to the placement of private funds through a lease-back type of arrangement. Each type of financing alternative, which are discussed in the paragraphs below, has its own risks and rewards.

TABLE 7-5

**YARD WASTE RECYCLING PROGRAM
PARTICIPATION AND DIVERSION RATES
COSTS AND REVENUES**

				Estimated %		$/Ton				
Location	Total Population Served	Total Households Served	% of Households Served	Participatio n (1)	Volume Reduction	Collection & Transport Costs	Processing Costs	Total Costs	Revenues	Net Cost
Davis, CA	44,000	10,000	100	70-80	50-60	NR	NR	NR	0	NR
Jacksonville, FL	700,000	300,000	100	80-90	80	NR	17.36	19.76	0.01	19.75
East Tavas, MI	2,600	1,350	100	70	65	10.00	<10.00	<20.00	0.00	<20.00
Montgomery County, MD	282,000	75,000	48	90-95	85	83.33	18.46	101.79	19.20	82.59
Wellesley, MA	27,000	8,500	100	90-95	60-65	0.00	11.11	11.11	0.50	10.61
Woodbury, NJ	2,329	825	17	80	70	43.00	15.00	58.00	0.00	58.00

SOURCES: Taylor and Kashmanian, 1988
HDR Engineering, 1993
Kelly, 1993

TABLE 7-6

INCREMENT COSTS OF YARD WASTE MULCH PRODUCTION PINELLAS COUNTY, FLORIDA

Cost Factor	Processing Costs $/Ton
Monitor for Rejects at Delivery	0.79
Tub Grinder[1]	9.71
Trommel Screen[2]	2.06
Windrow Formation and Turning To Meet State Guidelines for Safe Public Distribution[3]	7.92
Distribution[4]	0.90
Contractor Administrative	1.42
Equipment Contingency Plan	0.53
Reject Disposal[5]	0.75
Contractor Profits[6]	3.61
County Administrative	2.26
Total	$29.94

Notes:
(1) Includes all material handing costs in staging area (labor, equipment, O&M, capital, etc.)
(2) 11,000 Tons/yr., 37 percent of volume
(3) Material costs to build, turn, irrigate and remove (labor, equipment, O&M, capital costs)
(4) 2,500 Tons/yr., 10 percent volume
(5) Three percent Volume
(6) Amended at 15 percent

Source: Ragsdate, et al, 1992.

7.2.1 User Fees

In many jurisdictions, recycling programs are being funded through the imposition of user or uniform tipping fees on wastes delivered either to particular facilities (e.g., a landfill, transfer station, or waste-to-energy-facility) or to the community's entire solid waste stream. In this way, recycling programs become funded no different than operation of a transfer station fleet, provision of groundwater monitoring at a landfill, or undertaking of a household hazardous waste amnesty program. In the first case, only the users of a particular component of the system pay for the cost of recycling programs, while in the second case, all of the users of the community's system pay for these costs. A move towards these funding approaches has resulted from the imposition of full-cost accounting requirements in many states. The essential component of user fee systems is that they are operated on a self-supporting basis

"Full" cost accounting is a means of establishing the full direct and indirect costs of collecting and disposing of solid waste, picking up yard waste and recyclables, and providing all the necessary services such as costs of operation and maintenance of solid waste collection vehicles, costs of bond interest or loan payments, labor costs, fringe benefits, insurance, public information, billing services, special waste collection, etc. In many communities, this type of accounting program has been in existence for some time, particularly with rising costs of providing such services. It has been a means to establish good management methods to an essential municipal service and allowing the City fathers to compare the cost of municipal service against that which could be provided by private contractors. However, in many other communities these costs are unknown and are carried in different budget accounts in the community's financial management information system. In recent years, several states such as Florida, Georgia, and Indiana, among others, have enacted requirements for local governments to publish and file with the state their full and indirect costs of providing solid waste collection, recycling, and disposal services. The common objective in this legislation is to introduce rigor in financial management of solid waste systems by showing cost relationships in two main areas direct costs (wages, equipment, supplies, insurance, tipping fees, and closure funds, etc.) and indirect costs (management oversight, legal services, data processing and billing, etc.). The overall intent of this move away from tax-supported systems has been to shift the cost of an increasingly expensive solid waste management system to user groups.

Over the past few years, more than 200 communities have experimented with the use of volume-based, variable rate, or weight-based structures to pay for the cost of funding their recycling programs, as well as providing an incentive for waste reduction. There are many number of programs, but they can be classified into several major types: pre-paid bag, tag, or sticker systems; and variable rate can systems. In essence, all of these programs requires that solid waste collection customers select a subscription level based on individual needs. That is, the number of household refuse containers that is needed to dispose each week. Typically, the subscription levels offered are usually in terms of the standard 30-gallon container equivalents with some communities going to much lower container sizes such as 10 and 19-gallon sizes. The customer is then charged for each container left out on the curb.

The advocates for these type of programs point out that the volume-based rates have proven to be effective economic incentives to promote recycling. A number of studies undertaken by the City of Seattle, Washington have indicated that the introduction of a variable can rate helped the city slow the growth of solid waste tonnage requiring disposal in landfills, while also providing residents an incentive for recycling. Thus, the city was able to avoid implementing a mandatory recycling program, although weekly solid waste collection is mandatory for Seattle residents. Each household in Seattle is charged a monthly fee for garbage collection on a combined city utility bill (water, sewer, and garbage collection) based on the size of container used for weekly collection and if the container is collected at curbside or from the homeowner's backyard. Current monthly fees range from $11.50 per household for the smallest can (19 gallon) to $62.92 per month for a 90 gallon container collected from the resident's backyard. Based on these rates, the city has found that about 28 percent of its customers have signed up for the 19 gallon rate, 62 percent for the 32 gallon size, and only 10 percent for 60 and 90 gallon cart service.

While there are many clear benefits of volume-based rates to both the user and community, implementation of such systems may be hindered by institutional obstacles. Firstly, implementing volume-based rates requires that the community has the legal powers to establish solid waste rates either through its own billing system or to approve rates established under franchise or contract collection programs. Even if the community holds these powers or can establish an ordinance setting forth these powers, there have to be assurances that there are convenient recycling alternatives available to the public. This may require development of extensive public and private recycling programs along with public education and information efforts. All of these programs may require increased staff in areas where the community had not previously provided services such as customer service, billing, and public information. Community decision-makers also should not discount the need to develop complex financial/rate-making models and billing programs to help implement such volume-based systems. There also has to be some consideration of providing enforcement capability such as on-site inspections and recordkeeping. Lastly, there has to be consideration of providing assistance to citizens with fixed or low incomes. Volume-based rates may impose a burden on those income classes which may be somewhat mitigated by the enactment of special rates or providing economic assistance.

7.2.2 Property Taxes

Although many communities have established solid waste systems founded on an enterprise-based accounting program, property taxes continue to provide the basis of funding in many communities throughout the United States. Funding for solid waste services, including collection, transport, recycling, waste reduction, and recycling services can be usually found in the general revenue accounts of the community. This usually means that no separate billing or collection system is required thus making property taxes easy to administer. Additionally, many residents prefer this funding option because local property taxes are deductible on federal and many state tax returns.

There are, however, several major disadvantages to this type of funding. Unfortunately, this type of funding program does not accurately account for the "true" or "full" cost of providing solid waste services since many related municipal services (e.g., fleet maintenance, insurance, legal, accounting, etc.) are commonly left unaccounted. Thus, the cost of municipal solid waste collection and disposal programs under these circumstances have been traditionally considered to be "free" services by many residents and businesses, thereby hindering the development of effective recycling programs. Some have argued that only when residents and businesses are charged the real costs of providing solid waste services will recycling have a real chance of success. Furthermore, local property tax caps greatly restrict the overall level of funding for recycling programs since these programs must compete for limited public funds for other essential public services such as police, fire, education, and health.

7.2.3 Assessment Systems

A number of states currently allow local governments to fund the cost of their solid waste system operations through the imposition of special solid waste assessments on users. Typically, these assessments are placed on the non-ad valorem portion of the property tax bill on residences and/or commercial businesses within a specified service area, commonly called a municipal service benefit unit (MSBU). Billing for refuse service through a combined tax bill mechanism is less costly than if a separate tax bill would have to be sent out. Failure to pay these annual assessments are treated by many government jurisdictions to that of failure to pay property taxes. In some political jurisdictions, a lien is either placed on the property, which must be paid when the property is sold, or the lien is sold as a tax lien by the government entity third party. Very often, governments are hesitant to impose such conditions because individual homwowners who pay little or no property taxes may be forced out of their homes for failure to pay these non-ad valorem assessments. Some communities have addressed this problem by providing specific exemptions from paying these assessments by allowing senior citizen low-income discounts. This type of special assessment has gained favor in communities with limited waste stream control.

The assessment is established by the government entity to cover the cost of the service, in this case, solid waste collection, recycling, and/or disposal service. Generally, a flat fee is levied for the service regardless of the service level provided to all residential units and/or commercial businesses within the benefit unit. The benefit unit may consist of an entire franchise collection area or the entire governmental jurisdiction (e.g., incorporated or unincorporated areas). However, having separate benefit units for each franchise area could allow for different rates between different franchise areas.

However, assessment systems are not without their disadvantages. Commonly, a separate tax roll must be prepared which will include all properties within the benefit unit. Preparation of this tax roll may be somewhat costly, especially if there is poor cooperation between the government entity imposing the special assessment and the agency which formally issues the tax bill. Also, such assessment fee systems are often unpopular with local citizenry unless an effective public information program is developed which explains the need for the assessment program.

A good example of a special assessment program used to fund a local recycling program can be found in Lake County, Florida. In 1992, the County established a special assessment program in the unincorporated area of the county for billing the fee charged for disposal of solid waste, including the cost of the curbside recycling program, on the annual tax bill. Under this program, all residential property of four units or less per building (improved property) are assessed for solid waste collection and disposal services, with a credit against this special assessment given to those residential properties which subscribe to franchise hauler services and pay the special assessment through their franchise hauler billing. Exemptions from paying the special assessment are also provided to properties with a permit to haul solid waste which is generated on their property to a transfer station or drop-off facility. The special assessment is collected through the office of the Lake County Tax Collector.

7.2.4 Sales or Municipal Taxes

Sales taxes can provide significant funding for recycling programs, especially in areas which are heavily populated or have high tourism and recreational activities. Some states, like Florida and Georgia, allow communities to levy local option sales taxes on gasoline or other consumer goods and apply these sales tax revenues to the construction and/or operation of solid waste facilities. Generally, however, these levies are often inadequate for larger projects.

Many states also allow local communities, primarily municipalities, to levy taxes on utility services (telephone, electricity, cable TV, gas, water, etc.), utilized by residents in their community. By state statute the funds generated by these levies can then be used by the community for any lawful purpose. Again, as in the case of property taxes, these revenues are often used for essential public services which recycling services must compete against.

7.2.5 Bond Financing Options

Some recycling projects are capital intensive. Most communities do not have the available capital or tax base to "pay-as-you-go" during construction or procurement, so alternate financing mechanisms must be used. To the extent that a source of money is available to offset a portion of the construction cost, the amount of financing required can be correspondingly reduced.

The two primary expenses associated with financing are the interest rate on bonds or the expected internal rate of return for private funds and the cost of capitalized interest. The amount of capitalized interest is usually 35 percent to 40 percent of the bond size, and this will vary according to interest rates and the length of construction. The interest rate or rate of return is influenced by: (1) the bond market at the time of sale and whether bonds bear interest at fixed or variable rates; 2) whether the interest is tax exempt, and, if so, the extent of the federal tax-exemption; and (3) the security structure of the bond and credit quality of the obligations.

There are three financing options generally available for recycling projects: (1) grants; (2) bond sales; and (3) private funds. In addition, a combination of these options may be utilized. The following discussion focuses on bond sales and private funds because grant funds for construction of these projects are limited in most states. Bond structures most commonly used with respect to recycling facilities include: (1) general obligation bonds; and (2) revenue bonds. The most significant distinction between these alternatives is the nature of the security structure; in particular, the level of obligation assumed by the governmental entity.

7.2.5.1 General Obligation Bonds

General obligation bonds would pledge the full faith and credit of a community as security for payment of the bonds. The full faith and credit pledge includes the general revenue of the community which may include property tax, sales tax, license fees, income tax and other charges as payment for the bond principal and interest. The bond interest rate reflects the community's credit worthiness with less emphasis on project economics. General obligation bonds are generally tax exempt. However, if the bonds meet the criteria to qualify as private activity bonds, the tax-exemption is limited to taxes on the gross income of the bondholder. In addition, the bonds must meet additional criteria in order to qualify for even this limited exemption. In determining whether to use general obligation bonds, the following should be taken into consideration:

(1) Depending on the credit rating of the community, the interest expense may be lower than under the other alternatives discussed;

(2) The community will assume maximum financial exposure which may lead to downgrading of its credit;

(3) The obligations or bonds will impact the community's debt limit, and consequently reduce the community's borrowing capacity; and

(4) A public-vote will be required to approve sale of the bonds.

This type of financing has not been used by many communities to finance construction of its public works facilities and it is unlikely that general obligation bonds would be issued for a recycling project.

7.2.5.2 Revenue Bonds

Revenue bonds pledge the revenues available from the project as security for the bonds. The project revenues may include solid waste disposal fees, revenues from the sale of recovered materials, any available insurance proceeds, and damage payments from contractors, if applicable. Most revenue bonds are limited obligations, and as such, the community would not be obligated to raise taxes to pay debt service. However, the community will generally be required to covenant that rates charged for solid waste collection and disposal will generate revenues sufficient to cover all project costs including bond principal and interest payments. This "rate covenant" enhances the project's security and may help to lower interest rates. Another security feature is establishment of various bond reserve funds to provide a degree of protection against revenue shortfalls or unexpected operation and maintenance expenses.

Since project revenues are the security for revenue bonds, the bond interest rate will reflect how project risks are handled and the bond buyer's exposure to project risks. The bond holder's risk exposure is typically judged by bond rating agencies such as Standard & Poor's or Moody's.

Another type of revenue bond financing is often termed "Conduit Financing", although such financing is severely limited. Conduit financing, along with private ownership can be used in recycling projects when the tax benefits available to private corporations (as owners) offset project costs and, consequently, lower disposal fees. A private investor is considered eligible for certain tax benefits by contributing only a portion of the project costs from private funds with the balance of project funding supplied by tax-exempt, private activity bonds ("PABs") issued by a community on behalf of the private entity. The portion of the project costs contributed by the private owner is called the equity contribution.

The equity contribution will usually be provided during the construction period on a pro rata basis with other construction expenses. The private corporation may also provide a corporate guarantee for payment of the bonds, subject to certain conditions, as an additional security feature for the PABs. If an unconditional corporate guarantee is provided, the rating agencies will focus on the credit worthiness of the corporation as well as the project economics. However, in order to obtain an unconditional corporate guarantee, the corporation must believe that the project is an excellent risk.

As discussed above, in an equity participation arrangement, the contractor is also the project owner. Generally, the contractor is a limited subsidiary of a large parent corporation. The parent corporation must have a tax benefit appetite in order to take maximum advantage of available tax benefits. Because not all contractors can effectively use the available tax benefits, the transaction may be structured as a sale/lease back with a third party owner with a large tax benefit.

7.2.5.3 Tax Status of Private Activity Bonds

The bond interest rate will also be influenced by the tax-exempt status of the bonds. PABs are subject to many limitations which are not applicable to other bonds (which are not PABs) including restrictions on the type of facility financing and restrictions on the use of proceeds to pay issuance costs. PABs may not be refunded in advance, even if the criteria for tax-exemption are met. Interest on the bonds will be exempt from the gross income of the bondholder, but will remain subject to the alternative minimum tax on individual and corporations. As a result, PABs will bear higher interest rates than would non-PABs with comparable security features.

7.2.6 Private Funds

In addition to the revenue bond structure described above, there have been proposals for 100 percent private financing of recycling projects. Under this option, the private firm would raise the capital, design the recycling project, procure the equipment and/or construct the system, and operate the program. There are many options to this type of private alternative in terms of procurement, operations, and degree of ownership. However, to date sole private financing has not been evidenced as a preferred method of private ownership, and is not considered a viable option for most communities. There are several reasons why 100 percent private financing has not been implemented:

(1) The capital intensive nature of the project requires substantial corporate commitment. Investment of 100 percent of the project cost "ties up" capital which could be used in other projects:

(2) Tax benefits have been reduced in recent years and may be received for an investment of only a small portion of the project cost;

(3) Financial analysis of the corporation's costs of money and required rate of return on 100 percent rather than a small portion capital investment typically results in a higher operating fee requirement.

7.2.7 Bank Loans and Anticipation Notes

Many local governments have utilized bank loans for short-term project financing (rolling stock, site purchases, vehicles, etc.) since they do not have available capital out of their general fund to "pay-as-you-go" during the initial procurement of equipment or construction of the recycling facility. Bank loans are used until long-term bond debt can be issued under more favorable market conditions. These are termed, Bond or Tax Anticipation Notes, TAMS or BANS, respectively, and help communities to stabilize their cash flow. In some areas of the country, however, the interest rate on such loans can compare favorably with municipal bond rates, particularly since bond issuance costs are saved or deferred until a bond issue refunds the loan at a later date as the notes expire. Further, many local banks have long-term relationships within a community, thereby having a greater understanding of the local need and support for a project. This is often reflected in reducing the need for restrictive project covenants and complicated financial negotiations with parties distant from the community.

7.2.8 Leasing

Equipment leasing is prevalent in the solid waste industry, particularly with the acquisition of collection vehicles. Most often, an equipment manufacturer secures a leasing company purchases and holds title to the equipment during a specified period, commonly the usual length of depreciation of the asset (five to seven years). During this term, the community or private operator pays the leasing company a lease payment for use of this equipment. At the end of the lease agreement, the leasee can purchase the equipment at a value specified in the lease agreement. This type of lease arrangement is often favorable to the community since it defers the long-term capital funding of expensive, but depreciable, equipment.

Another form of equipment or facility leasing is often termed, "leveraged leasing". Under this type of leasing system, with the private entity (lessor) providing a portion of the capital needed to purchase the asset. The community then finances the balance of remaining capital needs with a financial intermediary who acquires the tax advantages of ownership. This allows this third party the right to depreciate the asset and perhaps be able to receive an investment tax credit, if one is available under federal law and certain state tax codes for recycling facilities and equipment. It is these tax advantages that allow the third party to receive a higher after-tax return on its investment which may be reflected in the lower interest on the cost of money for the community compared to other financing methods.

7.2.9 Grant Funds

Many states in recent years have funded local community recycling programs through program grants. Such funds have been used by communities to initiate the planning of recycling programs or they paid for the initial equipment acquisition for the community program. Other states have allowed grant funds to pay all or a major portion of the capital costs of facilities such as MRFs or the short-term operating costs of a curbside recycling program. Advanced disposal fees (ADFs) have been used by a number of states to fund grant programs which have then be used to finance various local government recycling programs. An ADF is a fee imposed on a product, such as tires, motor oil, white goods, batteries, and containers, which is intended to reflect its cost of disposal. By levying a fee on these consumer products, proponents argue that the consumer will then have an economic incentive to buy products which are more "environmentally friendly" and enhance their demand. However, research conducted by Arthur D. Little, Inc. has suggested that the low fees imposed in states which have enacted ADFs has little, if any, impact on reduction in demand for these products. Further, experience of states which have enacted ADF legislation has shown that ADFs are oftentimes difficult and costly to implement. Nevertheless, ADFs have been popular with many state legislatures since ADFs can provide a substantial funding source. In the case of Florida, for example, a recent ADF program on containers enacted in Florida projects an annual funding of solid waste grants in the range of $30 million.

A state recycling budget survey undertaken by Waste Age and Recycling Times in 1993 indicated that 46 states impose some form of user fee or tax as a source of funding for their statewide solid waste management program. These funds are used to support a variety of programs including support of state regulatory functions, community recycling programs, and market development. Despite cuts in some states in 1992 because of declining tipping fees revenues in states heavily hit by the recession, most planned to spend more than $200 million on recycling in 1993 with this figure expected to increase as the national economy picks up strength. Florida, Illinois, Michigan, Minnesota, Pennsylvania, and Wisconsin have the largest budgets for recycling grants to local communities. All of these states, for example, each provided more than $20 million a year in grants with Wisconsin and Pennsylvania providing grants totalling more than $36 million and $31 million, respectively.

7.3 Issuing and Marketing a Bond Issue

The implementation of complex public works projects as are many recycling programs requires the specialized technical, financial, and contracting skills that are often not found in local government. Consequently, public bodies contemplating these projects commonly acquire professional services to supplement in-house capabilities in such areas as engineering, finance, insurance, environmental science, and law. The financing of a recycling project involves these key participants beyond the principals to the transaction. The roles and responsibilities of each of these participants are discussed below.

7.3.1 Key Participants

7.3.1.1 Investment Banker

The investment banker or bond underwriter plays a principal role in the marketing of the bond issue. When such bonds are to be issued, the investment banking firm, either alone or with other partners, makes a proposal to the community for the purchase of its bonds, which are in turn sold by the investment bankers to investors. The investment banker's profit is derived from the spread between the price at which the bonds are purchased from the community and their ultimate resale to investors. Out of this spread, these firms must pay certain expenses associated with the bond issue such as: the fee of the underwriter's counsel, out-of-pocket expenses, and expenses incurred in connection with the delivery of the bonds.

The investment banker serves an important function for the community during the time in which bonds are being issued by assuming the risks of adverse price fluctuations. It is possible that the market price of these securities will decline before all of the issue is sold. As part of its fee, the investment banker underwrites this risk and ensures that the community will receive the amount of capital funds necessary for the recycling project.

7.3.1.2 Bond Counsel

The role of the bond counsel in public offerings is different from that of counsel to types of corporate financings. In public financings, the legal counsel essentially does not serve as attorney for any particular party, but acts as a sort of special counsel for the transaction itself. That is, the main function of the bond counsel is to render an opinion regarding the securities to be issued by the public body to pay for the construction of the project. As expected, the contents of this opinion vary from project to project. Essentially, however, the bond counsel is responsible for reviewing the bonds themselves as well as the official governmental documents, certificates, legal agreements, laws, decisions, and rulings authorizing their issue to ensure that the bonds were issued in a legal manner and constitute valid and bindings obligations of the community. If the bonds to be issued are to be secured by project revenues, the bond counsel's opinion will normally include a statement regarding the legality of pledging such revenue sources according to federal, state, and local regulations. Additionally, where the interest on the bonds are to be exempt from taxation, the bond counsel will be called upon to render a legal opinion as to the tax-exempt status of the interest on the bonds.

In addition to the duties outlined above, the bond counsel retained by the community may be called upon to prepare the necessary financing documents, such as the trust indenture and bond resolution, in connection with bond issuance and to advise on other matters related to the financing of the project.

7.3.1.3 Financial Advisor

Although financial advisors can be hired by any of the parties in a recycling project, they are usually employed by the community to review the financing plan and to advise the community on bond marketability and the expected interest rate based on the financing plan.

7.3.1.4 Independent Consulting Engineer

When the project is ready to be financed, an independent consulting engineer is hired by either the issuer or the underwriter of the bonds to prepare a feasibility report on the project. The independent consulting engineer is charged with reviewing the preliminary engineering plans, specifications, guarantees, and cost estimates for the project. Generally, the independent consulting engineer, whose report is included in the official statement for the bond issue, must make a determination about the project in the following areas:

- Is the project technically feasible?;
- Were the construction and operating cost estimates prepared using sound engineering estimating methods?;
- Are the estimates of the amounts of revenues produced by the project reasonable?; and
- Are the terms of the design, construction, and operations agreements in the best interest of the project?

7.3.1.5 Rating Agency

Bonds are assigned quality ratings that reflect their relative investment qualities. There are two major national bond rating agencies (Moody's Investor's Service and Standard and Poor's Corporation). These private bond rating agencies charge fees ranging from $2,500 to $50,000 or more for their time and expense to rate prospective bond issues. Generally, both of these rating agencies rate the issue considering whether the community has an acceptable economic base to repay the issue (income levels and growth, employment mix, etc.); whether their exists a good relationship between expenditures and revenues; what is the level of the total debt burden relative to the community's budgetary resources and taxbase; and does the community have the political and administrative leadership to carry out its functions for the project. Answers to these and other questions enable the rating agencies to place a rating on the proposed issue. Some of this information will be gathered from documentation provided by the community while other information will be gathered through face-to-face presentations. Bond ratings are not always needed for every project and many smaller communities may be unable to obtain a rating prior to obtaining capital for a project.

Bonds are assigned quality ratings that reflect their relative investment qualities for individual or institutional investors. Bonds with a rating of triple A (AAA) have the highest rating and are felt to be extremely safe, while double A (AA) and single A (A) bonds are often held in conservative investment portfolios. Triple B (BBB) bonds are considered to be investment grade bonds and as such are permitted by law to be held by banks and other institutions in most cases. Such bond ratings are important to both the issuer and the investor for many reasons. First, a bond's rating is an indicator of its risk. Thus, a bond with a high rating usually results in a lower interest rate than one with a low rating. Second, most bonds are purchased by institutions, which are restricted by law in many areas of the nation to investment grade issues. Consequently, if a bond receives a rating below triple B, the community may have a difficult time selling these bonds.

Prior to a bond offering, the community or its representatives presents to the rating agencies all of the pertinent documents related to the project's financing such as the indenture, consulting engineer's feasibility report, construction and operation agreements, and municipal bond insurance policy, if any. Analysis of these documents by the rating agencies forms the basis of their initial or conditional rating. There is no guarantee that such ratings will continue for a given period of time or that they will not be lowered or withdrawn entirely by the rating agencies if in their judgement circumstances so warrant.

7.3.1.6 Bond Insurance Company

To increase the marketability of their bonds, communities often purchase municipal bond insurance policies which are offered from groups of private insurance companies. Purchase of this type of insurance policy enables the community to receive a Triple A rating on the insured bond issue from both Moody's and Standard and Poor's. This triple A rating generally results in a lower interest rate on the bonds, thereby saving the community significant borrowing costs over the term of the issue. The insured bonds offer security to the potential bondholder and require that the insurance company to continue to pay the investors their coupon rate should the community default on their responsibilities to pay debt service. The insurance companies offering such policies have eligibility tests similar to those of the national bond rating agencies. The premiums charged by these companies typically range from a half to one percent of the total debt service due over the life of the bonds. Market conditions affecting the spread between insured and uninsured bonds must be evaluated, however, to determine whether there are significant net cost savings for each particular bond issue given the added cost of the bond insurance premium.

7.3.1.7 Trustee

A trustee is a trust company or bank with trust powers, authorized to do business within a particular state, which performs the functions defined in the bond indenture document. This Indenture of Trust or Trust Indenture is a legally binding agreement between the community and the trustee in which the community pledges its interest in the project as security for the payment of principal and interest on the bonds issued for the construction and operation of the project. The trustee may be responsible for the following tasks: dispensing the construction fund; directing the flow of revenues from the project; investment of monies in various funds as set forth in the Indenture; monitoring the levels of these different funds; paying the principal and interest payments to the bondholders; and for seeing that the bondholder's interests are protected during the terms of the bonds.

7.3.2 Steps in Bringing the Bond Issue to Market

Having noted the principal participants responsible for financing a recycling facility through a bond issue, discussion in the paragraphs below is a simplified listing of the general steps and documents necessary to bring the bond issue to market.

Step 1 Adoption of a Bond Resolution and Trust Indenture Bonds issued by a community are secured by contracts and legal documents which must be enacted by the governing body of the community. A bond resolution sets forth the maximum size of the issue; officially authorizes the various financing techniques which may be required to successfully market the bond issue (e.g., bond insurance, letters of credit, discounts, call provisions, etc.); authorizes the execution of an Indenture of Trust between the community and a trustee; outlines the procedure for obtaining interim borrowing; and authorizes the commencement of bond validation proceedings.

As noted previously, a Trust Indenture is a contract between the community and a trustee which contains provisions establishing adequate financial reserves for protecting the bondholders should problems occur in the project. It also specifies how the proceeds of the bond issue and the revenues generated by the project during its operation will be invested; how obligations to the contractor and bondholders will be paid; how much the trustee will be compensated; the circumstances under which the trustee can be removed; procedures for the modification and amendment to the Indenture; and the discharge of the Indenture. To carry out these major tasks, the Trust Indenture will typically provide for the creation of special funds such as debt service, revenues, operating and maintenance, reserve and contingency, and renewal and replacement.

Step 2 Validate the Bonds Local courts in most communities have jurisdiction to determine the validation of bonds. Any government entity may file a complaint in such courts where the proceeds of the bond issue are to be expended to determine its authority to incur bonded debt. This complaint is required to include information relative to the bond ordinance, resolution, or other legal proceeding authorizing the issue; the size of the issue and the amount of interest on each bond; and other pertinent matters for the court to weigh in its deliberations. The date for the hearing is required to be published well in advance of the hearing.

By requiring the legal notice, all citizens or taxpayers of the local area are made parties to the proceeding as if they were personally served. Consequently, any citizen or taxpayer can intervene in the validation proceeding. If in the final judgement of the court validates the bonds, and no appeal is taken within the prescribed time, then the judgement is forever conclusive, and the validity of the bonds can never be called into question in any court by any person or party. However, if any person is dissatisfied with this final judgement then he can usually appeal to the state's highest court within the time and manner prescribed in the state's appellate rules.

Step 3 Preparation of a Preliminary Official Statement An official statement is a document issued by the investment bankers or underwriters to the community in connection with the issuance of the bonds. The official statement will typically contain an introductory statement describing key financial and other features of the issue, and is followed by sections which summarize the pertinent financing documents and agreements, including the bond resolution, trust indenture, construction and operation contracts, market agreements, lease, or other contracts with third parties. The official statement will have certain appendices, such as the independent consulting engineer's report, audited financial statements of the community, and the opinion of the bond counsel as to the tax-exempt status of the bonds.

Generally, the official statement is preceded by a preliminary official statement which is issued primarily for marketing purposes and before the pricing of the bonds. Such documents are used by many investment banking houses to establish pre-sale interest in the issue and identify potential buyers among insurance companies, banks, investment bankers, municipal bond funds, pension funds, and others who may be interested in purchasing the community's bonds. Although the final official statement normally follows the preliminary official statement except for minor changes, the later document has a statement printed in red along the left margin of the front cover admonishing the purchasers of the bonds that they should only rely on statements in the final official statement. This document is commonly referred to as the "red herring".

Step 4 Meetings With Rating Agencies and Bond Insurance Firms Bond ratings are key to the successful financing of a recycling project. As indicated previously, both national rating agencies realize that each recycling project is a unique project, with unique security features and project risks. Consequently, the analysis of each project is tailored to that project's particular legal and economic structure. The community's investment bankers will at this point assist in providing all information necessary to these various rating agencies so that the bond issue can be properly evaluated.

Commonly, this information is presented through visits to the rating agencies by project personnel. This allows the community to more fully portray the need for its project and outline the managerial and financial strengths of the community.

Step 5 Blue Sky and Legal Investment Surveys "Blue-sky" laws are those state laws regulating the sales of securities. Such laws were passed by most states in the early part of this century to prevent unscrupulous security salesmen from operating in their cities and communities. The term "blue-sky laws" can be traced to a member of the Kansas legislature who remarked that such promoters would sell stock in the "blue-sky" itself. Today, most states have enacted "blue-sky" and investment laws. The "blue-sky" laws typically set forth the procedure under which new security issues and dealers or brokers of such securities are registered with the state. Many states also have enacted investment laws which regulate the security portfolios of financial institutions such as savings banks, insurance companies, trustees, and local governments. Since these regulations may affect the marketing of a bond issue, it is important that a survey of the different state laws be done before the bonds are issued. This survey is usually completed by the underwriter's counsel at the time the preliminary official statements are mailed to prospective investors.

Step 6 Establishing the Final Pricing of the Issue Prior to the submittal of a bond purchase contract to the community, the investment bankers arrive at a final offering price of the bonds based upon the results of their pre-sale distribution efforts. At the several pricing meetings held during this period, members of the investment banking syndicate, which were selected by the community for this bond issue, compare their expectations about capital market trends, the supply of competing bond issues on the financing calendar, secondary market activity, the size of the bond issue, and the buying patterns of investors.

Step 7 Submission of the Purchase Contract to the Issuer Commonly, the timing of the purchase proposal is usually determined by the investment banking syndicate. The investment bankers will negotiate the purchase of the bonds required for the project. At the time this proposal is submitted to the community, the investment bankers must provide the financial advisor with interest rates, spread, management fees and other information to assure the community that the bonds are being sold at the best possible price given the municipal bond market at that time. Based on an analysis of this information, the community can either accept or reject this bond purchase proposal. If this proposal is rejected, the community and the investment bankers commonly continue to negotiate in good-faith until an agreement is reached as to the purchase price, interest rates, and other terms or conditions of the sale of the bonds. At such time the community and the investment bankers reach agreement on these issues, a purchase contract is signed by both parties binding the investment bankers to purchase the bonds at a state price once the par amount of the bonds are delivered to them by the community.

The signing of the purchase proposal sets in motion several important activities. Commonly, it is at this time that the final official statement is printed and then distributed by the managing underwriters to other members of the underwriting syndicate, institutional and retail investors, and the community. Also, the managing underwriter coordinates the allocation of the bonds among the syndicate members and actively monitors on a day-to-day basis the actual selling of the bonds to investors.

Step 8 Bond Issue Closing On the day of the closing, the community delivers to the investment bankers its legally executed and binding, definitive bonds. Documents relative to the bond issue are reviewed by the community's bond counsel and the underwriter's counsel to be certain everything is in order. Simultaneously with the delivery by the community of its duly executed bonds, the investment bankers deposit with the trustee the full purchase price of the bond issue, as specified in the purchase proposal and the trust indenture.

-8-

INSTITUTIONAL ISSUES

This section will discuss some of the key institutional issues which must be addressed by a community in deciding to proceed. These include project procurement, financing, ownership, risk, and solid waste flow control considerations.

8.1 Procurement

In general, procurement encompasses the acquisition by a community of equipment and operational services. Procurement can be discussed in terms of approach and procedure. The procurement approach dictates the manner by which engineering, design, construction, start-up and operation are acquired and how responsibilities will be assigned between the public and private sectors. The procurement procedure determines the method by which such services are acquired and the legal guidelines which must be followed. The following discussion identifies the approaches and procedures generally available, the relationship to various risk assignments, and an analysis of the appropriateness of the available procurement alternatives.

8.1.1 Procurement Approaches

There are essentially three approaches used in procurement: (i) conventional or architect/engineer (A/E); (ii) turnkey; and (iii) full service.

8.1.1.1 Conventional

The conventional or A/E approach is the traditional and most widely used approach for procuring public works projects. A professional engineering firm is retained by a community to participate in the planning and design of the project. The engineer, acting as an agent for the community, prepares equipment and system specifications to be let for public bidding, and is responsible for certain design elements of the project. Following bid evaluation, the same, or a different, engineer is retained by the community to monitor construction of the project in order to ensure the use of proper materials, supplies, equipment, etc. Upon completion of the construction, the engineer assists in plant start-up and testing, and may be required to prepare operating manuals for the project. Once the project has passed acceptance testing, operational responsibility becomes that of the community which might either operate the project itself or contract for its operation with a private firm. This process usually requires several different contracts between the community and its engineering consultants, contractors and vendors.

MRF projects have recently incorporated certain aspects of full service

MRF projects have recently incorporated certain aspects of full service procurement in the A/E approach. Instead of bidding on individual pieces of equipment, the entire equipment process line is bid as a package. This minimizes the number of contractors which the community's construction manager must deal with and provides a mechanism for sharing the risk of project performance with the contractor. Design of the remaining "wrap-around" facilities is the responsibility of the engineer and is secured under a general contractor bid.

8.1.1.2 Turnkey

In a turnkey approach, a single entity is awarded a contract for design, construction, equipment, and start-up of the project. The turnkey contractor selects the equipment and supplies to be used and may either design and construct the project itself or subcontract portions of the work. In either case, the contractor assumes sole responsibility for construction of the project. Upon completion of construction, start-up and successful testing, the recycling facility is accepted by the community which then assumes responsibility for operation or contracts for the operation with a private firm. A modified turnkey approach may include a short term (e.g., one to five year) agreement with the contractor to operate the project after acceptance.

8.1.1.3 Full Service

An extension of the turnkey approach is to assign total responsibility for MRF design, construction, equipment, acceptance testing, operation, and possibly ownership to a single entity or full service contractor. Under this approach, the community is provided with long-term operation and maintenance services. This procedure usually includes a contract for design, construction and acceptance testing and another long-term contract for project operation and maintenance.

Table 8-1 summarizes the project procurement and operation responsibilities associated with each of the procurement approaches.

8.1.2 Procurement Procedures

The procurement procedure should provide a fair and equitable method of selecting a contractor which promotes public confidence and at the same time takes advantage of the competition which exists in the materials recovery industry in order to get the most beneficial arrangement for the community. Competition between contractors should not be limited to construction price alone, but should also take into consideration the quality and reliability of the MRF system proposed, performance guarantees and long-term operational and maintenance expenses as appropriate. The procurement should take these elements into consideration when the community evaluates various proposals prior to selection of a contractor.

8.1.2.1 Competitive Negotiations

The use of the competitive negotiation method of procurement is appropriate for those situations where the item or service desired requires extensive discussions with offers to determine the fairness and reasonableness of offers and to establish the allocation of risks upon which priced proposals will be made. In the public works area, "competitive negotiation" is better known as the "RFP process".

Procedures available for procurement of public facilities or services are generally subject to statutory bidding requirements or local ordinances. There are four generally acceptable bidding procedures: (i) competitive sealed bidding; (ii) competitive simultaneous negotiations prior to issuance of the final request for proposals; (iii) two-step competitive negotiations after issuance of the final request for proposals; and (iv) sole source negotiations.

8.1.2.2 Competitive Sealed Bidding

Competitive sealed bidding is the most common method used for acquiring supplies, services, and construction for public works projects. This method requires the preparation and issuance of an Invitation for Bids ("IFB") containing detailed specifications and a purchase description of the desired item, service or construction project. Upon receipt of bids, the community determines if the item or service being offered satisfies and is responsive to the requirements of the IFB and if the bidder is responsible. No change in bids is permitted once they have been opened. Once bid evaluation is completed, the award is made solely on an objective basis to the lowest responsive and responsible bidder.

TABLE 8-1

TYPICAL MRF PROCUREMENT AND OPERATION RESPONSIBILITIES

	Procurement Approaches		
Project Responsibilities	A/E	Turnkey	Full Service
Planning	C or E	C or E	C or E
Plant Design	E	V	V
Preparation and Issuance of Plant Specification	E	V	V
Construction Supervision	E	V	V
Construction and Equipment Installation	V	V	V
Start-Up	C or V	C or V	V
Operation	C or V	C or V	V
Ownership	C	C	C or V

Notes: C: Community
V: Vendor or Contractor
E: Engineer as Agent for the Community

8.1.2.3 Two-Step Competitive Negotiation or Simultaneous Negotiations

Under this procedure, a Request for Proposals (RFP), as opposed to an IFB, is issued. The RFP contains minimum technical and financial qualifying criteria, general system and performance specifications, the evaluation criteria to be used (price being only one) and the relative importance of each evaluation component, as well as the basic features of the risk allocation package including, in many instances a draft contract. Financial and minimum technical qualifications of bidders can also be performed in a separate Request for Qualifications (RFQ) issued prior to the RFP. Based on the RFQ responses received, a short list of qualifying vendors is prepared. After receipt of the draft RFP and an appropriate period to receive initial vendor comments and requests for clarifications, discussions are held with all qualified vendors to arrive at a final RFP which is issued and upon which the vendors are required to submit bids without modifications. This method allows for "apples-to-apples" comparison or evaluation of submitted bids.

The selected vendors are ranked based upon the comparative evaluation of their respective proposals, and the community then enters into final negotiations with the top-ranked proposer, usually for a specified time period, to arrive at the most desirable transaction for the community. If negotiations are not satisfactorily concluded within that time period, then the community is able to terminate negotiations with the first vendor and commence negotiations with the second vendor. Furthermore, in the unlikely event that two or more of the top-ranked vendors' proposals are virtually the same, the community would reserve the right to enter into competitive simultaneous negotiations with those vendors. By limiting the negotiating period with the selected vendors, the competitive negotiation process enables the community to structure a competitive procurement on a cost effective basis in circumstances where the number of qualified vendors would dictate against use of the competitive simultaneous negotiation approach.

This process differs from competitive sealed bidding in two major respects. First, judgmental factors are used to determine not only compliance with the requirements of the RFP but also their impact on the evaluation of competitive proposals. The effect of this is that the quality of competing proposals may be compared and tradeoffs made between the price and quality of offers. Secondly, since comments and contract discussions occur after the issuance of the draft RFP, and changes in proposals may be made to arrive at a final RFP which is the basis for each vendor's offer, the final RFP reflects an overall transaction which is most responsive to the community's needs and provides a competitive environment for procurement of vendor proposals. Final award is made to the offerer whose proposal is most advantageous to the community based upon price and specified evaluation factors set forth in the RFP.

The two-step competitive negotiation procedure is used chiefly in situations where the complexity of the system or service desired requires the preparation of detailed specifications by the community. This procedure incorporates features of both the competitive sealed bidding and competitive negotiation methods. The community may limit the number of offerers through an RFQ process. In step one, the draft RFP is issued and discussions are conducted separately with the vendors based upon their initial comments to ensure complete understanding by the vendors of what the community requires in order to enable offerers to be more fully responsive to the final RFP. This step is similar to the competitive negotiations, up to the issuance of the final RFP upon which unpriced proposals are submitted. Proposals are then evaluated based upon criteria set forth in the RFP and the two or three vendors submitting the most responsive proposals are selected for the negations phase of this procurement process which is completed simultaneously.

8.1.2.4 Sole Source Negotiation

Sole source negotiation involves no competition and is used when the community determines that there is only one source for the desired supply, service, or construction project.

Not all procurement procedures are applicable for use with the available procurement approaches. The competitive sealed bidding procedure is only appropriate for the A/E procurement approach since only in the A/E approach is the project broken down into well defined, discrete components of equipment, fabrication, on-site construction, and operation. Similarly, the essence of successful turnkey or full service procurement involves, in addition to considerations of price, a careful analysis of vendor experience, technology, guarantees, financial well being and the overall risk allocation reflected in the transaction. Such factors are subjective to some degree and the community must be permitted to make qualitative judgments as are provided in the competitive simultaneous negotiation, two-step competitive negotiation and sole source negotiation procedures. Table 8-2 summarizes the relationship between procurement approaches and procedures.

8.1.3 Procurement and Risk Posture

The basic difference between the available procurement approaches is the level of control the community can retain in the design, construction and operation of a recycling facility and the subsequent allocation of risk. A detailed risk assessment is provided in Section 8.4, and the following paragraphs provide a brief summary of key risk issues.

TABLE 8-2

APPLICABILITY OF PROCUREMENT APPROACHES AND PROCEDURES

	Approaches		
Procedures	A/E	Turnkey	Full Service
Competitive Sealed Bidding	Appropriate	Not Appropriate	Not Appropriate
Competitive Negotiation/RFP Process	Possible	Appropriate	Appropriate
Two-Step Competitive Negotiation	Not Appropriate	Appropriate	Appropriate
Sole Source Negotiation	Possible	Appropriate	Appropriate

Full service procurement allows the greatest sharing of risk with the private sector. In a full service procurement, contractor guarantees are available for a maximum construction price, completion with a specified construction period, demonstration and maintenance of specified performance levels, and a fixed annual operating and maintenance expense, usually subject to a negotiated escalation formula. A turnkey procurement contractor will typically guarantee a maximum construction price, completion within a specified construction period, demonstration of specified performance levels prior to the community's acceptance, and a limited warranty (generally one year) on workmanship and materials. In the A/E procurement approach, the risks which can be shared include completion within a specified construction period and demonstration of specified performance levels.

In each procurement approach, the guarantees are subject to occurrences beyond the contractor's control such as changes in law and force majeure events. Depending on the final negotiated agreement, there may be a maximum limit on the contractor's aggregate contractual liability.

There is a direct correlation between the amount of risk shared with the private sector vendor and the overall price of the project. Since a full service contractor typically takes the greatest amount of risk, a full-service proposal will be higher priced than the A/E procurement approach. Because the guarantees associated with the risk posture accepted by the vendor are conditioned on certain contract terms, it is essential that the procurement procedure also clearly defines these conditions and the community's risk position.

8.1.4 Analysis of Procurement Options

Determination of the procurement approach and procedure is one of the first decisions the community must make in the implementation. In making the procurement decision, the community must first establish what its goals are in implementing the project control, ownership and risk allocation structure.

The community's position on project control is influenced by: (i) traditional community practices in procurement of public work projects; (ii) the public's perception of community involvement in an environmentally sensitive issue, and (iii) the technical, managerial and financial capabilities of the community. The community should evaluate its position in response to several questions:

- Does the community want to own the project?
- Does the community to operate the project?
- Does the community have the technical and managerial resources needed to operate the project in an efficient manner?

- What level of involvement does the community desire in project design, technology selection, site layout, and architectural standards, etc.?

The A/E approach is particularly applicable if the community desires to own and operate the project. The approach allows the greatest degree of community control regarding design features and therefore, offers the greatest response to public concerns which may arise. However, the A/E approach also requires the greatest commitment of community resources in development and operation of the project and exposes the community to potentially greater economic risks. The key construction risk concerns are cost overruns, lack of sole source responsibility and potential technical failure. These risks are manageable and can be mitigated by selecting a competent and experienced A/E firm and selecting only proven technologies and equipment. In addition, a good construction manager can minimize scheduling and coordination problems.

Operational risks pose a more significant problem in the A/E approach. If proven and reliable technology and equipment is used, operational performance risks are minimized. However, with public operation of the project, a highly qualified technical staff must be maintained in order to assure that adequate maintenance and proper operation is performed. For communities which have similar public work activities, this does not pose a significant problem and the project staff can be formed relying in part on previous experiences. An additional concern is the continuity of financial commitment to maintain an adequate staff, particularly in light of fluctuations in the community's economy and budgetary constraints, as well as potential changes in political leadership's philosophy or support.

The turnkey procurement approach would be applicable if the community desires to establish sole source responsibility for project construction, yet wishes to keep the option of community operation of the project. The community relinquishes control of project construction once the plant specifications are determined, except for the cost of work change orders. However, because the community is responsible for providing long term operation either through community staff or contracted operators, it is in the community's best interest to monitor the design and construction. This additional monitoring expense, coupled with the profit margin for accepting sole source construction responsibility, usually makes the turnkey approach more costly than the A/E approach.

The key risk area in the turnkey approach is the long term performance of the recycling facility. This risk can be mitigated in part by selecting only sound and proven technologies and financially secure contractors. However, because the contractor is not responsible for operations, his profit lies in efficient construction, potentially resulting in a higher proposed capital cost, and there is some risk that construction short-cuts could result in operational problems over the long term or in additional expenses which are not apparent when the plant is tested and accepted. To partially alleviate this concern, a modified turnkey approach has been proposed which includes a short operating term (i.e., one to five years) by the contractor. However, this is still only considered to be a short term solution and due to the lack of substantial short term operating experience in the U.S. MRF industry, the effectiveness of such a modified approach has not been demonstrated. If the community desires to operate the project, the same concerns described in the A/E approach apply. If the community elects to contract for operation of the facility with a company other than the turnkey construction contractor, only limited operational risks would be accepted by such an operator due to the operator's lack of control over the design and construction. Furthermore, disputes arising over the cause of operating problems may be difficult to ascertain and resolve.

The full service procurement approach is applicable if the community views operation of the recycling program as obtaining a service, similar to the services the community currently receives for waste disposal. The degree of control available to the community depends, to a great extent, on the ownership status of the project. If the public maintains ownership, the community can require the operator, as the community's agent, to respond to the public requests on operating standards or procedures. However, if the community establishes more control on operating procedures than is considered to be standard in the industry, the full service operator will take fewer operating risks.

If the community elects to allow private ownership of the project, the public's control is limited to predetermined contractual rights such as annual inspections, maintenance of specific permit criteria, and other similar areas. Response to public concerns would depend on the working relationship established between the community and the private owner, and the private owner may not be legally required to respond to all problems. With private ownership, the community must also recognize that the ultimate responsibility for waste disposal rests with the community in the event of vendor bankruptcy, insolvency or shutdown.

In terms of risk allocation, the full service procurement approach offers the greatest opportunity for the community to share construction and operational risks with the private sector. Because one company has responsibility for all aspects of the project; design, construction, and operation, the vendor is willing to guarantee certain performance levels during operation, an annual operating and maintenance fee, usually subject to an escalation formula, and material floor prices. These operational performance and cost guarantees provide the community a more balanced risk allocation posture than is normally available for A/E and Turnkey projects. However, the full service contractor may also expect to be paid a higher operating fee or have a greater profit expectation for assuming such risks. There are certain risks beyond the contractor's control (e.g., changes in law, force majeure events, etc.) which will still be assumed by the community regardless of the ownership of the facility. The chief concern regarding the full service approach is the long term financial stability of the contractor.

Recently, new corporate sponsors with access to proprietary resource technology have entered the recycling industry. To protect the public's interests, a vendor with both a sound and proven technology, and the financial strength to backstop construction, performance and operating guarantees, must be selected. The community must also establish a contractual right to conduct regular inspections and reviews of the facility, and in the event the full service operator's corporate sponsorship varies, to maintain an independent financial guarantee of the parent company and its assignees and well as a separate contractual right to assure the continued availability of any proprietary technology.

A summary of the advantages and disadvantages of the procurement options is provided in Table 8-3.

8.3 Ownership Issues

Ownership of a recycling project is another important community policy decisions. Although it is generally considered to be a public responsibility to protect public health and provide for the safe disposal of solid waste, there are several reasons why private ownership of a recycling project is attractive to various communities:

(1) Communities which have relinquished the responsibility for solid waste collection and disposal to the private sector are also inclined to relinquish ownership and control of a recycling project;

(2) Private sector businesses (haulers) who either control solid waste flow through collection and/or disposal operations, can develop a project as part of their continuing business venture;

(3) Legal restriction (e.g., special labor regulations, bidding laws, contractual restrictions, etc.) or lack of cooperation between multiple communities make it too expensive or impossible to maintain public ownership; and

(4) Federal tax benefits are available to the private sector and may be applied to subsidize the cost of the project.

TABLE 8-3

ADVANTAGES AND DISADVANTAGES OF PROCUREMENT APPROACHES

Approach	Advantages	Disadvantages
A/E	• Potential lower price • Allows community control and input on design	• No single source responsibility • Coordination of multiple contractors • Higher community risk exposure • No long term performance guarantees
Turnkey	• Sole source responsibility for design, construction and initial performance	• No long term performance guarantees • Operating expense risk remains with the community • Community controls operation
Full Service	• Sole source responsibility for design, construction and operation • Greater risk sharing potential • Guaranteed long term performance	• Higher costs associated with long term operating risk • Reduced community control

The choice of public or private ownership is not always clearly apparent, and community decision makers must analyze various project aspects before making a decision. The principal areas to evaluate in making the ownership decision are: (i) economics; (ii) residual value; (iii) project control; and (iv) project risk and financing security. The following discussion focuses on the differences inherent in public and private ownership in these four areas.

8.3.1 Aspects of Public and Private Ownership Decision Making

8.3.1.1 Economics

Assuming that the cost of construction and operation of recycling project is the same regardless of ownership, the principal reason for the attractiveness of private ownership has been the availability to use federal tax benefits to subsidize project costs. The use of equity contributions will reduce the annual debt service requirement, either through an initial contribution and smaller bond size or through annual contributions during the initial years of project operation. A portion of the private owner's return on this investment is from the available tax benefits. This is taken into consideration when the tipping fee for the project is set by the private owner. Computer models are available to estimate the vendor's return and to optimize the equity contribution. There is no assurance, however, that the community would be given full credit for the available tax benefits and the worth of tax benefits to individual owner is difficult to determine during negotiations. In addition, an annual operating profit would be included as part of the operating and maintenance expense. Such an operating profit would be common to both public and private ownership if a private company operates the project.

The Tax Reform Act has significantly reduced the tax benefits available to private owners. Prior to the Tax Reform Act, the available tax benefits included a 10 percent investment tax credit (ITC) and accelerated depreciation (5 years for the majority of depreciable items). These benefits had allowed the owner to provide an equity contribution of 20-25 percent of the project costs. Under the revised tax code, the ITC has been eliminated and the depreciation period extended. In addition to lowering the available tax benefits, the Tax Reform Act also lowered the maximum corporate tax rate to 34 percent resulting in a reduction in the owner's tax benefit.

8.3.1.2 Residual Value

The term of a bond sale or service agreement is typically 10 to 20 years; however, the useful life of a project (with proper repair and maintenance) may be in excess of this term. If properly maintained with replacement of parts on a regular schedule, the residual value of the project may be the same as the original construction price. When making the ownership decision, the community should take into consideration:

(1) Barring any major technological breakthrough, the project would still offer an effective materials recovery system;

(2) The community would, in all probability, continue to rely heavily on operation of the project as the primary means of commercial waste recovery; and

(3) For continued use of a privately owned project after the initial term of the service agreement, the community would be required to either renegotiate the service agreement or purchase the project.

8.3.1.3 Control

Solid waste projects are particularly vulnerable to public scrutiny regarding "environmental concerns" and some public officials prefer to have more extensive control over the operation of a project than is afforded by private ownership in order to satisfy these public concerns. With public ownership, the community has control over all aspects of project construction and operation. Some public officials prefer to distance themselves from public involvement in such projects and prefer private ownership. However, as previously discussed, the community could only have limited control over operation of a privately owned project. This typically includes only the rights to inspect the plant and require periodic tests to demonstrate guaranteed performance levels.

It becomes a subjective decision for the community to weigh the public's reaction to project control when making the ownership decision.

8.3.1.4 Risk and Financing Security

The relationship between procurement options and risk was discussed in Section 8.1.3. With regards to ownership, the risk posture under a full service procurement approach is basically the same with either public ownership or private ownership. In private ownership, however, the owner may be willing to assume a portion of the risk of certain unforeseen situations. The owner's risk position is generally limited to providing additional capital in the same proportion as its initial equity contribution and is generally limited in amount to the available tax benefits. In addition, in a private ownership arrangement, the community may be asked to take the risk of loss of anticipated tax benefits due to changes in tax laws (i.e., tax indemnification).

In terms of financing security, public ownership backed by a general obligation pledge would provide the most desirable investor security arrangement. If revenue bonds are used under either public or private ownership, the vendor may be willing to supply a corporate guarantee or letter of credit (LOC) support as additional security enhancement for the bonds. The credit worthiness of the corporation and the extent in which limited subsidiaries are used are factors in the investor's security analysis.

8.4 Risk Assessment

Risk is the possibility that an event will occur which has detrimental impact on the project. A risk assessment is an evaluation of possible risk events, the impact of a risk occurrence, and methods by which risks can be reduced/mitigated or shared/allocated. The result of a risk assessment is a definition of the community's risk and a determination of the implementation strategies which best reflect the community's risk posture. the community risk posture is defined in the contractual arrangements, procurement approach and financing structure selected for a project. The following sections provide the community with an overview of project risks and propose a risk posture for consideration by the community. The community's actual risk posture should be determined once a "go/no go" decision on project implementation is made.

8.4.1 Risk Events, Impacts and Allocation

Risk events can be categorized in the following general areas: (i) waste stream; (ii) project construction; (ii) project operation; (iv) materials market; and (v) financial and contractual matters. Risk events can be further categorized as to the cause of the event (i.e., community-caused, contractor-caused or uncontrollable events) and appropriate mitigation or risk sharing measures taken. It is difficult to assess the probability of the occurrence of a risk event. However, events which are caused by the action or inaction of the community or caused by circumstances beyond the contractor's reasonable control will be the responsibility of the community. Events caused by or in the direct control of the contractor will be the responsibility of the contractor. Table 8-4 outlines typical risk sharing under the various procurement options.

The results or impacts of a risk occurrence are primarily financial in nature, and can include one or a combination of the following situations:

(1) Cost overruns and increased capital requirements;

(2) Inability to use the project for commercial recycling and subsequent decrease in landfill capacity due to continued landfilling or increased nonprocessed waste or residue with increased landfilling requirements;

(3) Increased operating and/or maintenance costs;

(4) Lost or lower than anticipated materials revenues;

(5) Lost or lower than anticipated tipping fees;

(6) Temporary or permanent MRF shutdown; and

(7) Failure to make debt service payments.

The probability that a risk event will occur can be reduced or mitigated by:

(1) Selecting only proven and reliable technologies;

(2) Selecting creditworthy contractors with demonstrated performance and management skills;

(3) Proper planning for project implementation, such as selecting project design and operating criteria to compensate for variations in waste composition and supply, providing screening measures for hazardous waste, providing protection for explosions, floods and other force majeure events, making provisions in the design for additional air emission equipment, etc.;

(4) Legislative control of waste deliveries;

(5) Establishing reserve funds for major repairs; and

(6) Securing reasonable insurance policies.

The degree to which risks can be shared by the community with a private contractor depends on the type of procurement approach selected. Project risks will be accepted by private contractors; however, adequate compensation for such risk sharing will be expected.

The community's risk posture should take into account that in order for any risk sharing arrangement to work:

(1) The risk allocations must be equitable to all parties;

(2) A good working relationship between the community and the vendor must be established;

TABLE 8-4
ALTERNATE PROCUREMENT APPROACHES RISK SHARING[1]

			FULL SERVICE	
RISK ELEMENT	A/E	TURNKEY	PUBLIC OWNERSHIP	PRIVATE OWNERSHIP
Waste Stream				
Waste Availability				
Failure to Deliver	Community[2]	Community	Community	Community
Haulers Strike	Community	Community	Community	Community
Competitive Diversion	Community	Community	Community	Either
Flow Control Failure	Community	Community	Community	Community
Waste Composition				
Evolutionary Change	Community	Community	Either[2]	Either[2]
Change in Law	Community	Community	Community	Community
Project Construction				
Contractor Fault				
Design Errors	Community	Vendor	Vendor	Vendor
Construction Strike	Community	Community	Community	Either
Subcontractor Failure	Shared[2]	Vendor	Vendor	Vendor
Cost Underestimated	Shared[2]	Vendor	Vendor	Vendor
Equipment Performance Failure	Vendor	Vendor	Vendor	Vendor
Uncontrollable Circumstances				
Subsurface Conditions	Community	Community	Community	Community
Force Majeure	Community	Community	Community	Community
Change in Law	Community	Community	Community	Community/Shared
Inflation Increases	Either	Either	Either	Either[3]
Tax Law Change	-----	-----	-----	Shared[2]
Project Operation				
Contractor Fault				
Excessive Downtime	Community	Vendor	Vendor	Vendor
Damage by Waste	Community	Shared	Shared	Shared
Equipment Underperformance	Community	Vendor	Vendor	Vendor
Uncontrollable Circumstances				
High O&M Cost	Community	Either	Either	Either[2]
High Inflation	Community	Community	Community	Either[2]
Change in Law/Regulation	Community	Community	Community	Community
Lower Material Revenues	Community	Shared	Shared	Either

NOTES:
(1) Community fault events are always considered community responsibility. "Either" reflects community or vendor responsibility; "Shared" reflects partial community and vendor responsibility.
(2) These fault events are usually resolved by negotiations.
(3) The risks of construction cost increases during the operation period may be handled through a fixed price contract (Vendor risk) or escalating payment (community risk).

(3) Natural positive incentives (e.g., revenue sharing) should be used to ensure performance; and

(4) Flexibility should be provided in long-term operating agreements so that a contractor can maintain a minimum profit during hardships thus avoiding bankruptcy or insolvency.

8.5 Waste Flow Considerations

One of the more critical issues facing public officials pursuing recycling facilities is what is commonly termed "flow control". In essence, a community must be able to assure those who will be operating a MRF and the financial underwriters for such a project that the recyclables generated from commercial and industrial establishments within the community will be available on a long-term basis. Without strong control of the recyclable waste stream, there is the potential for diversion of recyclables from the community's facility. This would be an unacceptable situation because the revenues from tipping fees and the sale of recovered materials are used to finance the construction and long-term operation of such facilities.

As consequence, flow control has been an issue of controversy in recent years between local governments on one hand, and the private waste recycling industry and waste haulers on the other hand. This latter group has argued against the imposition of monopolistic waste flow control by local government since it believed this would severely restrict continued access to recyclable materials taken from the waste stream.

This argument has been rejected by many communities and the investment community. Spokesmen for these groups have argued that the financing of recycling projects cannot take place without the long-term assurance on the part of government that a community's recyclables is committed for delivery to the project. Without such assurance, the investment community has asserted that the interest rate for project financing would increase dramatically. Furthermore, representatives of local government have asserted their rights to account for, collect, and process these recovered materials and the ownership of these materials for final disposition. They argue that the only way they can achieve recycling goals is to control the collection and processing of recyclables. Some communities in recent years have attempted to take a middle course by enacting waste flow ordinances with commitment for residential curbside recycling, while at the same time encouraging the development of a strong private recycling industry in their community.

In order to develop integrated programs, local government must consider all solid waste generated within its jurisdictions. It must also consider that waste which is imported or exported and its impact on the performance of the integrated system. Your community needs to determine and facilitate a mix of public and private initiatives that will make up the local system. Without the authority to control the fate of waste within the jurisdiction, governments will not be able to assemble long lasting, economically viable recycling and waste management systems.

8.5.1 Flow Control Mechanisms

8.5.1.1 Waste Flow Control Through Legislation/Regulation

A community can exercise some type of legal or regulatory authority over the collection, removal, and disposal of solid waste within its jurisdiction. Courts have long upheld the right of governments to adopt reasonable regulations in this area since all property rights are considered to be superseded by local government's police powers. Most of the court cases involving solid waste were decided by jurists at the turn of the century on the premise that regulatory authority was essential to public health and safety, since without such control solid waste would become a nuisance to neighboring property owners.

The waste flow control ordinances for solid waste facilities must be carefully drafted by local government. They must balance the needs of government to assure secure waste supplies for its facility against the legitimate economic concerns of collectors and the recycling industry to remove recyclable materials from the waste stream. Many local governments, while enacting strong flow control ordinances, now permit the recovery of recyclable materials from the waste stream.

8.5.1.2 Contractual Control of Waste Stream

Rather than resorting to the enactment of waste stream control legislation, a community can assure adequate quantities of solid waste for a recycling facility through contractual controls. This is accomplished when the community enters into long term contracts with other municipalities and private collectors to deliver recyclables to a recycling facility. This method of voluntary contractual commitments can be particularly effective to secure an adequate core of recyclables for the facility.

8.5.1.3 Economic Incentives for Waste Stream Control

Waste stream control can also be achieved by a community through economic incentives. For example, the operator of the recycling facility can attract recyclables from both public and private collectors by charging a zero or lower tipping fee than alternate disposal methods such as the community landfill. In this case, private haulers would be attracted to the facility since they would have no economic incentive to dispose of their recyclables at less convenient sanitary landfills elsewhere.

In order to accomplish this type of economic control over recyclables for a recycling facility, a community must be willing to subsidize the loss of project revenues with funds from some other source, such as from the general fund, a user fee, a tax or grant funds. For example, a user fee for solid waste can be established for different residential, commercial and industrial accounts whereby the proceeds from this fee could be used to subsidize the tipping fee at the facility. Some communities have also used revenue from property or other local government taxes to subsidize the tipping fee at their facilities. Use of property taxes, however, may be viewed by the investment community as a general obligation of the community and could result in a lowering of its bond rating.

8.5.2 Recent Court Decisions

On June 1, 1992, the U.S. Supreme Court rendered decisions on two of the most far-reaching cases involving solid waste in recent years. In Fort Gratiot Sanitary Landfill v. Michigan DNR, and Chemical Waste Management v. Hunt, Governor of Alabama, the Court ruled that states have no right to raise barriers to waste from out of their region. In both cases, the states had imposed either limits for disposal or regulations prohibiting the disposal of out-of-state waste. The Court's removal of these barriers now appears to pose major difficulties for solid waste flow control within states as well. Consequently, if haulers find cheaper rates across borders than inside their own region, they may haul the waste out of the local region, potentially bankrupting in-county facilities. Thus, communities, which set tipping fees based on expected waste flow, may be unable to guarantee that their waste will arrive at the facility to be processed.

A recent example of such a situation that exists in Mecklenburg County, North Carolina (Charlotte Area). Pursuant to state law, the community enacted a Solid Waste Management Plan, approved by the state, which required implementation of a flow control ordinance requiring all waste generated within the community to be disposed of at a community operated or licensed solid waste facility or facilities under contractual arrangements with the community. Pursuant to its Solid Waste Plan, the community entered into a long-term contract with Browning Ferris Industries, Inc. (BFI) for disposal of almost all of its solid waste stream at their private landfill. The community then charged a blended tipping fee rate at its designated solid waste facilities (waste-to-energy facility and BFI landfill) located in Cabarrus County, North Carolina, to recover its total net costs after energy and recycling revenues. In January 1992, a flow control ordinance was adopted by the community requiring all licensed community haulers to deliver their wastes to these sites.

Subsequently, Container Corporation of Carolina (CCC), a Fort Mill, South Carolina solid waste hauler licensed by Mecklenburg County, was informed by the community that its license would be suspended because of CCC's violation of the community's solid waste flow control ordinance. CCC had disposed of the solid waste collected on its community routes in a permitted private landfill in South Carolina. Having received the notice of violation from the community, CCC challenged enforcement of the community's flow control ordinance in U.S. District Court claiming that the community ordinance violated the Commerce Clause of the United States Constitution. On August 11, 1992, the Court issued a permanent restraining order enjoining the community from suspending CCC's flow control ordinance. The community has indicated that it will appeal this decision. In the interim, BFI has suspended its contract with Mecklenburg County because the county could no longer guarantee promised solid waste tonnages. County commissioners are currently considering an interim step of raising property tax rates to pay for this anticipated revenue shortfall of $1 million a year in tipping fees. The county also is considering delaying implementation of its COM-MRF project until this flow control issue is resolved by the Federal District Court and issues its final decision.

In May 1993, the U.S. Supreme Court agreed to review a challenge to a local flow control ordinance in New York (C&A Carbone, Inc. v. Clarkstown 182 A.D. 2d 213, 587 N.Y. S. 2d 681) which had been upheld by New York's highest court. Clarkstown's flow control ordinance requires that all waste originating in the town, as well as all out-of-town waste processed in the town, must be disposed of at a designated disposal facility. However, recyclable materials may be sent out of town for processing. The hauler, Carbone, is currently challenging the flow control ordinance contending that it has effectively put it out of business. An expansive decision by the Supreme Court striking down the Clarkstown ordinance could have severe impacts on community's attempting to finance the construction of a disposal or recycling facility. Notwithstanding the decision of the Supreme Court, the issue of flow control will continue to play a major role in the way recycling facilities are sized and recycling programs are developed, financed, constructed, and operated for the foreseeable future.

-9-

CASE STUDIES

Over the past decade, hundreds of communities across the United States have implemented many different kinds of recycling programs. Some have concentrated on the residential sector by emphasizing curbside collection programs for single-family homeowners. Others have attempted to minimize their capital expenditures for new recycling collection vehicles through the use of drop-off centers and blue bag collection containers. Still other communities have concentrated on diverting yard and woody waste materials from their landfills to mulching or composting facilities.

As we have shown throughout this book, there is no magic formula for success in a recycling program. Each community must balance its recyclling goals against available resources (time, money, and manpower), the cost and availability of alternative solid waste reduction and disposal systems, and institutional obstacles. We believe that the case studies discussed in the pages which follow illustrate creative programs whose decision-makers had to balance these factors against the light of political reality. These represent just a small sampling of the many excellent programs which are currently in operation on a day-to-day basis throughout the United States.

The State of Delaware

The 660,000 people living in the two rural and one more densely populated counties that make up the State of Delaware benefit from recycling and other solid waste management services provided state-wide by the Delaware Solid Waste Authority. As part of the State-wide Solid Waste Management Plan, residents are served by modern landfills, transfer, material and energy recovery facilities as well as a municipal waste compost plant. Prior to adopting its "igloo based" drop-off recycling program, the Authority's Delaware Reclamation Plant (DRP) was processing 1,000 tons per day of waste to recover 35% of New Castle County's (population 442,000) municipal waste stream.

In 1990, Delaware residents, most of which unaware of recycling already occurring via the DRP, made it clear to local leadership that a recycling program was of great interest. After investigation by the Authority, curbside collection on a statewide basis was not found to be feasible. Subsequent to the Authority's investigations, the General Assembly passed the Waste Reduction and Recycling Act, which directed the Authority to add on a comprehensive statewide system for source-separated material recycling. The Act provided no funding for the program, which lead to the Authority adding a $2.00 per ton surcharge in fiscal year 1992 on the user fee charged at the Solid Waste Management Centers.

Having had a positive experience with recycling igloos in a glass collection program with the Glass Packaging Institute in Kent and Sussex Counties, the Authority began the "Recycle Delaware" program, utilizing similar igloos, color-coded for the collection of specific materials. Browning-Ferris Industries (BFI) was selected by the Authority as the contractor for the program, responsible for the collection and marketing of the materials.

As of September, 1991, the Authority had successfully sited and opened 100 drop-off centers, 15 months ahead of schedule. The Authority attributed the rapid development schedule to a high degree of public interest and its dedicated staff.

Centers were placed in response to interest, following a solicitation letter to schools, fire companies, cities, towns, shopping centers, businesses employing over 200 persons, and other establishments that would accommodate a Center. Initial apprehension on the part of potential sponsors was diminished as it became clear that the Centers were attractive and kept clean.

Once specific sites were accepted, the Authority paved each site, placed the igloos along with proper signage and provided detailed information to the sponsor of the Center. After consideration of

Population: ***660,000*** **Annual MSW Tonnage:** ***785,000***

Type of System: ***Voluntary residential sort, drop-off centers, mixed waste processing and composting.***

Materials Collected: ***Glass and plastic bottles, bi-metal cans, aluminum cans, old newsprint, magazines, used oil and household batteries.***

Percentage of Waste Recycled: ***18%*** **Landfilled:** ***82%***

Tipping Fee: ***$58.50***

Source: ***N.C. Vasuki, Chief Executive Officer***
Delaware Solid Waste Authority
P.O. Box 455
Dover, Delaware 19903

The State of Delaware (Cont'd)

potential markets, the Authority selected color separated glass, newspaper and magazines, plastic bottles, aluminum cans and steel cans for inclusion in the drop-off program. The igloos are color coded to match specific recyclable products. In addition to recyclable materials, the Authority collects household batteries in a separate box and stored in plastic drums. Lithium batteries are deactivated and stored separately. Thirty-four sites throughout the State have oil collection igloos and provisions for oil containers that cannot be re-used.

Some unwanted materials appear at the drop-off sites. They often include plastic bags and injection-mold containers like margarine tubs, yogurt containers, aluminum foil, junk mail, and in some instances, worn tires and lead-acid batteries. The Authority has placed major emphasis on keeping the sites clean and free from debris which encourages community pride and support for the Centers.

Public education includes extensive outreach work done over the years by the Authority in association with its overall waste management mission. Radio and print messages reinforce the link between community volunteer efforts and the success of the drop-off program.

Having reached a consistent number of operating Centers, the Authority is currently reviewing operations and costs, and evaluating the actual performance of each Center. Centers that have low volume and high operating costs, have been moved to new locations in phases. The Authority maintains a master waiting list for sites that have been offered. Alternates are developed as the need to improve specific Center performance arises.

Cost per pound to collect the materials from the Centers and transport them to market average slightly under seven cents per pound ($129 per ton). The Authority is looking at measures which would further reduce the cost per pound.

City of Tampa, Florida

In the early 1980's, the City of Tampa implemented the McKay Bay Refuse-to-Energy Project, which was a rehabilitation and expansion of the City's former incinerator, to incorporate energy recovery and incineration capacity of 1,000 tons of municipal solid waste per day.

The City of Tampa began its solid waste recycling program in 1988 shortly after the State of Florida enacted comprehensive waste reduction and recycling legislation. This legislation was recently amended in 1993 and provides for a 30% waste recycling or reduction goal by the end of 1994.

At the outset of the City's recycling program, two pilot curbside recyclables and yard waste collection programs were implemented. These programs proved to be successful and were expanded in 1990 from 14,643 singlé-family homes to some 26,020 single-family homes in 3,000 multi-family units, and the curbside yard waste program servicing approximately 19,000 homes. Newspaper, glass containers and aluminum cans are collected in curbside recyclables program under contract to Waste Aid Systems, Inc. at a cost of $1.70/home/month with a 50/50 revenue split. This program has been averaging about 380 tons of recyclables per month.

The City's yard waste collection program includes curbside collection from about 40% of the homes in the City, as well as collection from a City brush site, illegal dump sites and special collection. Yard waste is collected by municipally operated routes with materials brought to the City's contractor, Baycycle Limited Partnership. Yard waste is then processed into mulch and some further into compost at a cost to the City currently set at $22.63 per ton. This program averages about 1,000 tons per month.

In addition to curbside collection, the City has a single buy-back and 23 neighborhood drop-off centers. The City entered into a contract with the Tampa Housing Authority and its Tenant Association to construct and operate the buy-back center employing local labor.

The Center began operation in March 1993 and currently purchases cardboard, newspaper, glass, scrap aluminum, and office paper. The drop-off centers are located at neighborhood parks, supermarkets, schools, and churches where glass, aluminum, newspaper, and plastic PET and HDPI bottles. The drop-off centers allow residents to recycle plastics since they are not collected in its curbside program. On the basis of lessons learned in a pilot program, City staff developed procedures to efficiently collect recyclable materials at its drop-off centers, keeping costs down and sites clean, and

Population: ***288,565*** **Annual MSW Tonnage:** ***350,000***

Type of System: ***Curbside sort collection with additional yard waste pickup; drop off centers; buyback center***

Materials Collected: ***Newspaper, aluminum cans, ferrous metals, glass, plastics, Christmas trees, telephone books, household batteries***

Percentage of Waste Recycled: ***10%*** **Landfilled:** ***10%*** **Incinerated:** ***80%***

Tipping Fee: ***$65.00***

Source: ***Barbara Kropf, Recycling Coorinator***
Solid Waste Department
4010 West Spruce Street
Tampa, Florida 33607

City of Tampa, Florida (Cont'd)

emptying containers before they overflow.

The City Council has enacted an ordinance requiring city agencies to purchase recycled products. During 1992 the City purchased more than $1 million of supplies with recycled content. Further, the City has been a major sponsor of an annual seminar in the Tampa Bay area encouraging businesses to buy recycled products and supplies.

Orange County, Florida

Since the early 1980's, the community has studied the feasibility of various solid waste disposal alternatives to sanitary landfilling, including development of a waste-to-energy facility, composting and recycling programs. These studies culminated in the completion of a comprehensive Refuse Disposal Facilities Plan which was adopted by the Board of County Commissioners in January, 1990. This Plan details the County's solid waste management goals, needs, necessary facilities, and the timing of key County decisions through the year 2015. As part of this planning effort, the Board of County Commissioners adopted earlier a Comprehensive Recycling Element of the Refuse Facilities Plan on February 20, 1989. The County's goal is to use recycling and composting programs, in conjunction with private recycling efforts, to reduce by 30 percent the amount of waste landfilled by the end of 1994. In recognition of its rapidly increased solid waste tonnages and these new regulations, the County has initiated several recycling pilot studies over the past few years to prudently plan for existing and future solid waste growth.

The existing Countywide recycling program consists of three phases. Phase I was implementation of a curbside residential recycling program for single-family homes. Starting in 1990, almost 200,000 households, both in the incorporated and unincorporated portions of the County are currently receiving this service. This system has been quite successful with a recycling rate of almost 30 pounds per household per month.

Recyclable materials are transported directly to the Recycle America MRF, which is located at the Landfill, or from the MRF Transfer Station, which provides convenient service to haulers collecting recyclables in the residential curbside recycling program. At the MRF, the recyclables are processed and marketed. The County MRF is operated by Recycle America under an initial five-year term contract with the County. Approximately 65 percent of single family households in unincorporated Orange County and participating cities utilize the County's vendor for processing and sale of recyclables. Phase II of the County MRF contract allows the existing MRF to be expanded a minimum of 300 tpd and includes an automated mixed recyclables processing line. The County's MRF processed a total of 20,255 tons of recyclable materials during its first 12 months of operations (August 1990 - July 1991).

The second phase of the recycling program involves yard waste collection and composting. A Countywide yard waste recycling program was initiated in 1991. This program includes two parts: (1) a "Lay it on the Lawn" program to encourage residents to leave their grass clippings in the yard; and (2) a yard waste collection and composting program. Starting January 2, 1992, the County and most cities began a similar program to collect yard waste curbside in the residential areas of the County.

Population: ***701,292*** **Annual MSW Tonnage:** ***1,486,560***

Type of System: ***Curbside sort collection with voluntary participation. Materials recovery facility for processing.***

Materials Collected: ***Newspaper, aluminum cans, ferrous metals, glass, plastic***

Percentage of Waste Recycled: ***20%*** **Landfilled:** ***80%***

Landfill Tipping Fee: ***$30.65***

Source: ***Stanley Keely, P.E., Deputy Director***
Division of Public Utilities
109 East Church Street
Orlando, Florida 32801

Orange County, Florida (Cont'd)

Over the past two years, the County has studied multi-family and commercial recycling through the implementation of various pilot studies. As a consequence of these efforts the Board of County Commissioners in July, 1993 approved the implementation of a commercial recycling program requiring all commercial waste generators in the unincorporated areas to recycle certain materials. Municipalities are also required to establish commercial recycling programs. A "blue bag" based system will be available to all such generators for delivery of recyclables to the County's transfer stations or landfill. This program is anticipated to be operating by Spring, 1994.

Solid Waste Authority of Palm Beach County, Florida

Organized recycling programs in Palm Beach County date back to the 1970's when the City of Boca Raton initiated a curbside collection program for old newspaper. In 1987, the Solid Waste Authority of Palm Beach County (SWA) authorized the implementation of a countywide recycling program in cooperation with Palm Beach County's 37 cities and private industry. Its first pilot curbside program was begun in March 1988 with some 2,196 single family homes. Nine cities move ahead with operation of drop-off facilities during this same time period. During the next year and a half, the Authority added an additional 12,000 homes in 13 cities and two in the incorporated area.

In May 1988, an interim materials recovery facility was constructed at the Dyer Landfill to sort and further process residential recyclables. An intensive public education and information program was also begun at this time. These activities were funded by grants received by the Authority from the Florida Department of Environmental Protection. All of the curbside programs were serviced by SWA vehicles for an interim period of six to nine months.

During the past five years, the SWA has expanded its recycling program to include 280,000 single family homes receiving curbside service and 140,000 multi-family homes receiving containerized collection. This approximates about 75% of the County's total residential population. Trials of co-collection of garbage and recyclables, automated high-rise recycling, and rural collection programs are being evaluated.

With the re-bid of the solid waste collection franchises, the SWA is planning a major effort in commercial recycling in addition to its on going efforts in office paper collection and glass and cardboard recycling from bars and restaurants. An intensive waste auditing and business education program has been established.

In addition to its recycling program, the SWA has implemented a permanent 550 ton per day materials recovery facility, a pilot yard waste and waste water sludge composting facility, a 2,000 ton per day RDF-fired waste-to-energy facility, Class I and Class III landfills, five transfer stations, a household hazardous waste collection facility, and a vast array of public education and outreach programs.

Population: ***883,000*** **Annual MSW Tonnage:** ***1,354,026***

Type of System: ***Curbside sort; materials processing facility***

Materials Collected: ***Aluminum and ferrous cans, plastics, newspaper and glass***

Percentage of Waste Recycled: ***28%*** **Landfilled:** ***32%*** **Incinerated:** ***40%***

Landfill Tipping Fee: ***$43.00***

Source: ***Kathleen Kelley, Recycling Coordinator***
7501 North Jog Road
West Palm Beach, Florida 33412

City of Omaha, Nebraska

The City of Omaha implemented a Citywide curbside blue bag recycling program in 1990. An agreement was signed with Refuse Resource Recovery Systems (RRRS) to receive, separate non-processibles, and sort out recyclables from the City's residential and multi-family waste stream collected by the City's contract garbage collector. Residents were asked to separate out their recyclable materials in blue plastic bags and the remaining waste in green plastic bags. Once the waste was delivered, RRRS was responsible for processing the recyclables and transporting the residue to the County landfill. The City paid RRRS $9.00 per ton, which was approximately 80 percent of the County Landfill gate fee.

Once the City program was operating, RRRS was unable to remain profitable. In late 1990, RRRS notified the City that it needed a significant increase in tipping fees to remain in business and subsequently, RRRS ceased operation. The City then solicited bids for operation of the program and Omaha's City Council approved a contract with Waste Management of Nebraska, Inc.

Over the past two years, the City of Omaha and its recycling facility contractor, Waste Management, have had to reinstitute the curbside recycling program after RRRS's termination. Undoubtedly, this uncertainty has impacted the performance of the recycling program and the public's participation rate. However, the results obtained in Omaha, using the blue bag collection approach, generally can be said to mirror the positive results shown in Pittsburgh, Chicago, and other communities.

For the most part, the collection element of the City's blue bag program has remained unchanged. Blue bag pickup is for 100,000 single-family homes with no commercial or multi-family participation. Blue bags are available from grocery stores, either free as small plastic grocery bags or by purchase of special 13 and 30 gallon blue bags. The homeowner is asked to sort his recyclables into two blue bags (1.35 mil. thickness) for glass, aluminum cans, HDPE and PET bottles, bi-metal cans, and a separate bag for newspaper. Watts Trucking company collects both recyclables and non-recyclable waste in the same truck on each homeowner's regularly scheduled collection day (once per week pick-up) and delivers the material to the City's Recycling Center.

Waste Management of Nebraska (WMN) currently operates the City's Recycling Center under a multi-year contract (through 1995). Once RRRS was terminated by the City, WMN took over responsibility for the Center's operations, which is a combination transfer station and MRF. The new MRF operation commenced

Population: ***311,681*** **Annual MSW Tonnage:** ***128,016***

Type of System: ***Co-collection of blue bags and refuse; delivery to materials processing facility.***

Materials Collected: ***Film plastic, glass bottles, ferrous and aluminum cans, and newspaper.***

Percentage of Waste Recycled: ***3%*** **Landfilled:** ***97%***

Landfill Tipping Fee: ***$13.65***

Source: ***Louis J. Tomsu, III***
Public Works Department
5600 South 10th Street
Omaha, Nebraska 68108

City of Omaha, Nebraska (Cont'd)

March 25, 1991, and currently receives an average of 500 tons per day.

Current operations begin with dumping of mixed refuse (blue bags and refuse, yard waste) onto the tipping floor of the transfer station area.

These materials are then stockpiled and later moved by a front-end loader onto a metal pan conveyor. Workers then manually pull blue bags off this conveyor with household refuse, yard waste, and other bulky reject conveyed to a transfer trailer for delivery to the landfill. After passing this initial sorting operation, the blue bags are then conveyed to the sorting operation, which occurs on the building level above the tipping floor and is sized for a maximum of 80 tons per day. Here bags are first manually torn open by workers. Material is then transferred by belt conveyors across a series of picking stations and magnets where film plastic, glass bottles (by color), aluminum cans, HDPE and PET plastics, and bi-metal cans are retrieved. A trommel is used at the end of the conveyor line for screening fines from the ONP. Recovered ONP and plastics are baled and then shipped using box cars located in an adjacent rail spur. Glass is stored before shipment in "Gaylord" boxes. Aluminum and bi-metals cans are densified and baled. Plastics are baled and shipped by truck.

During 1991, a total of 3,630 tons of blue bag material was sorted of which 2,932 tons were recyclable. The other 698 tons was classified as residual or non-recyclable material. As a comparison, the City of Omaha landfilled a total of 124,386 tons during this same time period.

The City's contract with WMN has a complex fee structure with the following specified conditions.

- Receipt of co-collected solid waste and separation of blue bags = $5.74/ton

- Transportation to the Landfill of non-recyclable waste and residual from sorting process = $2.95/ton

- Current tip fee at the Landfill = $13.65/ton

- Processing recyclables in blue bags and from drop-off center = $97.18/ton

- Facility and administrative fees = $0.68/ton

- Revenues received from sale of recyclables at following guaranteed floor prices, 40 percent to WMN and 60 percent to the City, estimated at $0.99/ton:

ONP	$ 35
HDPE	100
PET	100
Aluminum	600
Steel Cans	40
Glass	65
OCC	20
Kraft paper	30
Blue bags	60

Based upon these contract conditions, the City of Omaha estimates that its 1992 costs to operate this program at $23.97 per ton.

Union County, New Jersey

Union County is located in the northeast portion of New Jersey and is home to 502,000 people. There are 21 municipalities within its 103 square miles. The Union County Utilities Authority is responsible for the management and disposition of more than 550,000 tons of New Jersey Type 1 waste (includes municipal solid waste, bulky waste, vegetative waste, animal and food waste and dry industrial waste) each year. The Authority has developed a solid waste management system that consists of:

County-wide curbside collection of source separated materials from residential dwellings; a yard waste composting program; bulky waste recycling; Used motor oil recycling; household special waste collection and disposal; household battery collection and disposal; commercial mixed waste recycling; a waste screening program (visual inspection of 58 waste delivery vehicles each week); and energy recovery from non-recycled materials at the Union County Resource Recovery Facility.

A combination of public and private collection is provided. The Authority is responsible for collection in 11 municipalities. Private vendors collect the balance and provide processing for all materials.

Since January 1988, the majority of the County's solid waste has been transported via two private and one public transfer stations to out-of-state landfills. As a result of the Authority's desire to reduce its reliance on out-of-county disposal facilities it has proceeded with the waste management system described above. As of January 1994, the entire system will be in place and operable.

Material designated for recycling in Union County include: clear glass; brown glass; green glass; newsprint; aluminum; ferrous and non-ferrous metals; wood waste; and yard waste.

The waste screening program is designed to discourage haulers from attempting to dispose of batteries, drywall, paints, tires, electronics and automotive materials from the residential, commercial, institutional and industrial sectors. These materials are precluded as part of a program to minimize heavy metals, sulfur, PCBs and nitrogen oxide (NOx) that result from the combustion process at the Resource Recovery Facility.

The Authority is striving to achieve the New Jersey State goal of 60% recycling (counting all New Jersey Type 1 Waste).

Population: ***502,000*** **Annual MSW Generated:** ***550,000***

Type of System: ***Mandatory curbside container collection and yard waste composting.***

Materials Collected: ***Glass,old newsprint, aluminum, ferrous and non-ferrous containers, yard waste, household batteries, used motor oil, household special waste, and wood waste.***

Percentage of Waste Recycled: ***35%*** **Landfilled:** ***65%***

Landfill Tipping Fee: ***$13.65***

Source: ***Jeffrey Callahan, Executive Director***
Union County Utilities Authority
24-52 Rahway
Elizabeth, New Jersey 07202

The Town of Islip, New York

The Town of Islip is one of the early developers of an integrated solid waste management systems. It consists of a materials recovery facility; a yard waste compost facility; a waste-to-energy plant; and a methane-to-energy facility located at its closed landfill.

In 1980, the Town of Islip initiated its voluntary recycling program titles, "Wednesday Recyclables are Picked-Up" or WRAP. This program was expanded town-wide in 1982 when the Islip Resource Recovery Agency was formed to, among other things, implement a mandatory recycling program. In 1987, the Town purchased and distributed 20-gallon recycling containers to each of the 75,000 residences and renamed the recycling program, "We Recycle America...and Proudly". Initially, paper, corrugated cardboard, glass bottles and ferrous and aluminum cans were picked up and transported to the original material recovery facility (MRF) for processing. In 1989, plastic bottles were added to the list.

The material recovery facility opened in 1990 is capable of processing up to 400 tons of recyclables per day. In its first year of operation it handles nearly 30,000 tons of material. Automated side loader package trucks and traditional rear loader packers are used to collect recyclables and deliver them to the facility each Wednesday. The facility features a drop off area for residents with used motor oil. After commingled recyclables are deposited on the tipping floor, a combination of mechanical and manual processing occurs. Separated materials are sold to local and out-of-state markets. Glass from the facility is used to make glassphalt for local paving operations.

The yardwaste composting facility was opened in 1988 and processes more than 50,000 tons of year waste per year. Located on a 40 acre site in Ronkonkoma, New York, the facility accepts leaves, grass clippings, tree and brush debris that is collected once each week in the Spring and Fall from town residences. A window approach to composting is employed. A final compost product is provided free of charge to Town residents who come to the facility. The material is also marketed to local landscaping and agricultural businesses. Demand for the product is high and orders to meet commercial demand is often difficult to meet.

The MacArthur Energy Recovery Facility was commissioned in 1990. It converts 520 tons per day of municipal solid waste to 12.5 megawatts of electricity (enough power to supply 8,000 homes). The plant was designed to be a zero-discharge facility, which equalizes and clarifies its wastewater for quenching the process ash. In addition to energy recovery, roughly 400 tons

Population: ***300,000*** **Annual Tons Generated:** ***250,000***

Type of System: ***Curbside sort, materials processing facility; yard waste composting***

Materials Collected: ***Aluminum and ferrous cans, glass, plastic bottles, newspaper, cardboard and glass***

Percentage of Waste Recycled: ***28%*** **Landfilled:** ***32%*** **Incinerated:** ***40%***

Landfill Tipping Fee: ***$85.00***

Source: ***Charles Weidner***
Islip Resource Recovery Agency
40 Nassau Avenue
Islip, New York 11751

The Town of Islip, New York (Cont'd)

of ferrous each month are recovered from process residue. The recovered ferrous is sold to an overseas market for $16 per ton.

The Town of Islip has an extensive public education program that focuses on children in grades kindergarten through twelve. The thrust of the program is a comprehensive recycling education curriculum distributed to all of the teachers (grades kindergarten through six) in the 56 schools within the 11 school districts in the town. In addition, town representatives visit the schools to discuss recycling with students in all grades.

Five examples of the town's public education efforts include:

* The "Wasteless Lunch" program for students in the elementary schools. Parents are urged to minimize waste by preparing lunches with less disposable packaging, such as replacing throwaway paper bags with a lunch box. The students are responsible for weighing and charting their newly packaged lunches and comparing them to totals from the previous week prior to implementation of the program to encourage them to waste less food. Throughout the lunch period, educational videotapes are playing in the cafeteria while town representatives are available to provide the children with answers to their questions;

* For older children, a "WRAP-rap" program was sponsored that encouraged teens to write rap songs about recycling. High school students crated rap songs that featured informative lyrics which highlighted the town's WRAP program. The winning group will have their 60 second rap song recorded and aired on local radio stations;

* A household battery recycling program that provided schools with 55 gallon recycling containers in which to dispose of old batteries (the batteries were taken to an approved disposal site by contracted haulers);

* An Adopt-A-Highway Program, operated through "Keep Islip Clean", an affiliate of Keep America Beautiful, Inc., was developed for 1,500 miles of town-controlled roads. Volunteer groups are responsible for making a two year commitment to clean up litter from a stretch of road at least four times per year. The areas of responsibility are marked with signs bearing the name of the group in recognition of their contribution to the community;

* A paper recycling program has been successfully implemented in five of the school districts. This program involved student participation and interaction with local waste paper dealers. Lists of dealers are distributed to the schools to assist them in choosing a buyer for the paper that is separated, collected and stored by the students.

Groups of children are also taken on tours of various facilities in the town to see what actually happens to their daily waste.

In addition, public education pamphlets, posters and flyers are distributed to constantly spread the recycling message to the residents of the community.

Beyond school based programs, the adult population is indirectly encouraged to participate via peer pressure. As the balance of the neighborhood places recycling bins at the curb each week, it is hard not to be drawn into "keeping up with the Jones".

Mecklenburg County, North Carolina

Mecklenburg County is located in southern North Carolina and has experienced rapid population growth over the past few decades. The City of Charlotte, its largest municipality, has become one of the major banking and distribution centers in the United States. Since the 1960's, Mecklenburg County has embarked on an integrated solid waste management plan emphasizing waste reduction, recycling, waste-to-energy, and sanitary landfilling of the remaining waste.

The County began its recycling program in 1977 with the establishment of an initial drop-off center at a local high school. Since that time the County has sited 17 additional centers throughout the county to serve all residents, but mainly those who do not have curbside recycling. Fifteen (15) of these centers are open seven days a week, 24 hours a day and are unstaffed. The centers each average about 18 tons per month, and the program collects about 3,000 tons annually. The two staffed centers are accessible on a scheduled basis.

The County implemented a pilot curbside recyclables collection program in 1987 involving 2,400 households. In 1990, this pilot was expanded citywide in Charlotte and to several neighboring cities. Currently, the 120,000 single-family homes in Mecklenburg County are served by curbside recycling. A single 14-gallon red container is provided to each homeowner. Collection of recyclables occurs weekly on the same day as curbside trash collection. Participation rates for curbside collection is estimated by the County to exceed 80% on a monthly basis. Further, on the average, each home that recycles is contributing 380 pounds of recyclables a year to the program. The County has entered into a contract with Fairfield County Redemption (FCR) to process and market these recyclables. FCR operates a 34,000 square foot building which processes a daily average of about 95 tons per day.

The County pays no fee to FCR with FCR earning revenues on the materials it sells.

Population: *511,433*　**Annual MSW Generated:** *719,752*

Type of System: *Curbside sort, drop-off centers, and yard waste collection*

Materials Collected: *Newspaper, kraft paper grocery bags, plastic bottles, steel and aluminum cans, glass, white office paper, yard waste, motor oil, vehicle batteries, cardboard, and scrap metal*

Percentage of Waste Recycled: *9%*　**Landfilled:** *80%*　**Incinerated:** *11%*

Landfill Tipping Fee: *$37.00 (Residential Assessment)*

Source: *Brenda Barger*
Source Reduction Program Manager
700 North Tryon
Charlotte, North Carolina 28202

Curbside collection for multi-family dwellings began in January 1993 with service made available initially to approximately 50 complexes. When the "phasing-in" process is complete, approximately 205 multi-family complexes will have curbside services.

Since a substantial portion of the County's waste stream is yard waste, early efforts were directed to yard and wood

Mecklenberg County, North Carolina (Cont'd)

waste recycling. These efforts began in 1983 and were greatly increased following Hurricane Hugo's arrival in September 1989. In January 1991 the City of Charlotte began collection of yard waste in clear plastic bags at curbside. In preparation for this program, the County opened two mulch/compost facilities where it is processed through a slow speed shredder and tub grinder. The product is then placed into windrows for biological treatment. A final process involved separating the product into two different size ranges (> ⅜" (mulch) and < ⅜" (compost). The County has contracted with a local company to market the material under the name ECO Products for sale in retail outlets.

The County also has active office paper and telephone book recycling programs. Plans are to extensively expand commercial recycling efforts.

The County is in the process of developing a comprehensive source reduction program for implementation. Program strategies will include intense targeted campaigns designed to satisfy the State's reduction goal of 40% by 2001.

Mecklenburg County also has an active waste-to-energy program. The 230 tons per day University City Resource Recovery Facility went on-line in 1989 and with a 600 ton per day facility in the final implementation phase.

City of Pittsburgh, Pennsylvania

The City of Pittsburgh is in itself a collection of diverse neighborhoods ranging from elegant older homes to high-rise communities and two and three family style homes situated on many winding and narrow streets. There is a large elderly population as well as a significant number of students attending local colleges and universities. Known primarily as a white collar business and academic center in the 90's, not long ago Pittsburgh was a major industrial center with the steel industry a dominant force. Given its recent effort in improving air and water quality, citizenry are proud of Pittsburgh's environmental evolution. Recycling has become a major focus of Pittsburgh media, as noted by the frequent advertisements shown on television and newspaper.

The City's mandatory was launched in September 1990. The entire program was implemented in three phases beginning with 52,000 residential units (single, multi-family and high-rise). As of September 1993, all 131,000 single residential units, 32,000 multi-family units and 2,512 major institutions in the city are included in the program. Commercial establishments are mandated to recycle corrugated, high-grade paper and aluminum. This requirement is enforced by recycling inspectors that cite commercial establishments for not collecting recyclables. Commercial pickup is by private haulers.

Pittsburgh's residents are asked to place the following recyclable materials into semi-transparent, blue plastic bags:

- Aluminum
- Ferrous
- Glass
- Newspapers (separate blue bag)
- Plastics (HDPE and PET)
- Tin and Bi-Metal Cans, including empty aerosol cans

The bags are available at no charge (8 gallon) at grocery store checkout lanes. The large grocery chains in the Pittsburgh area have been major proponents of the blue bag system. Commercial bags manufactured under the names "Glad" and "Hefty", and others are also available in larger sizes (13- and 30-gallon) for retail purchase (roughly $0.13 each).

Once each week, on the normal garbage collection day, the bagged recyclables are placed curbside along with normal mixed waste for collection in rear-loading packer trucks by City crews. Twenty-five (25) cubic yard packer trucks from the existing reserve fleet, which have been painted blue, collect bagged recyclables separately. Using existing compaction rates, each recycling truck is able to service about 1,100-1,400 units per route as compared to the normal 400 to 600 stops per day for garbage collection.

As a result of the "Blue Bag Program", the City of Pittsburgh has been able to recycle nearly three million pounds of material from its

Population: ***369,879*** **Annual MSW Tonnage:** ***289,000***

Type of System: ***Curbside using blue bag plastic container; delivery to materials processing facility.***

Materials Collected: ***Aluminum and steel cans, glass, plastic containers, newspapers, ferrous aerosols and paint cans.***

Percentage of Waste Recycled: ***11%*** **Landfilled:** ***89%***

Landfill Tipping Fee: ***$16.15***

Source: ***Maribeth S. Rizzuto, Recycling Coordinator***
917 City County Building
Pittsburgh, Pennsylvania 15219

City of Pittsburgh, Pennsylvania (Cont'd)

residential waste stream. At the time of our tour, city officials estimated that participation in the program has been high, with a daily set out of 50-60 percent and monthly participation of 80 percent or more. This equates to an average daily tonnage sent to the materials processing facility ranging from 55 to 75 tons per day.

The City currently funds all solid waste collection and disposal service, including its recycling program, through general property taxes. There are no user fees to fund operation of the system. The State of Pennsylvania has subsidized the cost of program development and public education for recycling of up to 90 percent of first time, although long-term performance. The City has received grants from the State for costs in the range of $5 per ton of recycled waste, which is funded by a surcharge of $2.50 per ton on all waste received at landfills and waste-to-energy facilities.

Prior to initiating what was referred to as the "Pittsburgh One Bag Program", the Mayor and the City administration were sharply criticized for venturing into unexplored territory and departing from the suburban model. However,the City felt it was unable to pay an estimated $3 million in up-front costs for rigid recycling bins and specialized collection trucks under a traditional program.

The City's recycling budget for 1992 was $2,184,000. This includes all of the City's 1992 costs, including: salaries, fringe benefits, operating expenses (vehicle costs, maintenance, and fuel), and overhead (19 percent) based on a recent indirect cost study.

GLOSSARY OF RECYCLING TERMS

AERATION The process of exposing bulk material, such as compost, to air.

AEROBIC A biochemical process or condition occurring in the presence of oxygen.

ANAEROBIC A biochemical process or condition occurring in the absence of oxygen.

BALER A machine used to compress recyclables into bundles to reduce volume.

BIODEGRADABLE MATERIALS Waste material which is capable of being broken down by microorganisms into simple, stable compounds such as carbon dioxide and water. Most organic wastes, such as food remains and paper are biodegradable.

BUY BACK CENTER A facility where individuals bring recyclables for payment.

COLLECTION SYSTEM The total process of collecting and transporting solid waste. It includes storage containers, collection crews, vehicles, equipment, and management and operating procedures.

COMMERCIAL WASTE Waste materials originating in wholesale, retail or service establishments such as office buildings, stores, markets, theaters, hotels, and warehouses.

COMPOST Relatively stable decomposed organic material; the result of the composting process.

COMPOSTING The controlled biological decomposition of organic solid wastc under aerobic conditions.

CONSTRUCTION AND DEMOLITION WASTE (C&D) Building materials waste, dredging materials, grubbing waste and rubble from construction, remodeling, repair or demolition of buildings, bridges, pavements and other structures.

COST SAVINGS The monetary savings realized through waste reduction and recycling as a result of avoiding landfill or other disposal processes; sometimes referred to as "avoided cost".

CULLET Clean, generally color sorted, crushed glass used to make new glass products.

DIVERSION RATE A measure of the amount of waste material being diverted for recycling compared with the total amount that was previously thrown away.

DROP-OFF CENTER A method of collecting recyclable or compostable materials in which the materials are taken by individuals to collection sites and deposited into designated containers.

ENTERPRISE FUND A fund for specific purpose that is self supporting from the revenue it generates.

FERROUS METALS Pertaining to, or derived from, iron; often used to refer to materials that can be removed from the waste stream by magnetic separation.

FLOW CONTROL A legal or economic means by which waste is directed to particular destinations.

FOOD WASTE Animal or vegetable wastes resulting from the handling, storage, sale, preparation, cooking and serving of foods.

FRONT-END SYSTEM A process for salvaging certain reusable materials from the waste before combustion or other processing.

FRONT-END LOADER A collection vehicle with arms that engage a detachable container, move it up over the cab, empty it into the vehicle's body, and return it to the ground.

FRONT-END RECOVERY Mechanical processing of as discarded solid wastes into separate constituents.

GARBAGE Spoiled or waste food that is thrown away, generally defined as wet food waste; although in common usage garbage refers to all materials that are discarded as unnecessary.

GENERATION The act or process of producing solid waste.

GLASS An inorganic product of fusion that has cooled to a rigid condition without crystallizing.

GRADE A term applied to a paper or pulp which is ranked on the basis of its use, appearance, quality, manufacturing history, raw materials, performance, or a combination of these factors.

GREEN WASTE A combination of non-animal food and yard waste collected and composted together.

HAMMERMILL A type of crusher used to break up waste materials into smaller pieces or particles, which operates by using rotating and failing heavy hammers.

HDPE High density polyethylene, a plastic resin used to make items such as plastic milk and detergent containers, and base cups for plastic soft drinks.

HIGH GRADE PAPER Relatively valuable types of paper such as computer printout, white ledger, and tab cards.

INDUSTRIAL WASTE Those waste materials generally discarded from industrial operations or derived from manufacturing processes.

INSTITUTIONAL WASTE Solid wastes generated by schools, hospitals, universities, museums, governments, and other institutions. Some communities define institutional solid waste as commercial solid waste.

INTEGRATED SOLID WASTE MANAGEMENT A practice of disposing of solid waste using several complementary components, such as waste reduction, recycling, composting, energy recovery, and landfilling.

IN-VESSEL COMPOSTING A composting method in which the compost is continuously and mechanically mixed and aerated in a large, contained area.

INTERMEDIATE PROCESSING CENTER A type of materials recovery facility (MRF) that processes residentially collected mixed recyclables into new products available for market; often used interchangeably with MRF. An acronym is IPC.

INVESTMENT TAX CREDIT A reduction in taxes permitted for the purchase and installation of specific types of equipment and other investments.

JUNK Old or scrap metals,rope, rags, batteries, paper, rubber, junked, dismantled or wrecked automobiles or parts thereof which are not held for sale for remelting purposes; unprocessed materials suitable for reuse or recycling, commonly referred to as secondary materials.

KRAFT PAPER A paper made predominantly from wood pulp produced by a modified sulfate pulping process. It is a comparatively coarse paper particularly noted for its strength; in unbleached grades is used primarily as a wrapper or packaging material.

LANDFILL A site for the controlled burial of solid waste according to federal and state rules and regulations.

MAGAZINE PAPER A variety of coated and uncoated papers used in magazines and similar periodicals.

MAGNETIC SEPARATION A system used to remove ferrous metals from other materials through the use of magnets.

MANDATORY RECYCLING Programs requiring by ordinance or statute that residents or businesses keep specific materials from their solid wastes.

MANILA PAPER Indicates color and finish and not the use of manila hemp.

MANUAL SEPARATION The separation of recyclable materials from waste by hand sorting.

MATERIALS RECOVERY FACILITY A common acronym is MRF.

MATERIALS RECOVERY The concept of resource recovery, emphasis is on separating and processing waste materials for beneficial use or reuse.

MAXIMUM RECYCLING POTENTIAL The maximum amount of recycling possible for a community given an ideal market, regulatory, citizen participation, and technological limits. MRP is sometimes used as an acronym.

MECHANICAL SEPARATION The separation of waste into various components using mechanical means, such as cyclones, trommels, and screens.

MIXED KRAFT BAGS Consists of baled used kraft bags free from twisted or woven stock and other similar objectionable materials.

MSW COMPOSTING Mixed or Municipal Solid Waste Composting, the controlled degradation of municipal solid waste including some form of presorting to remove non-compostable inorganic materials.

MULCH Ground or mixed yard wastes placed around plants to prevent evaporation of moisture and freezing of roots and to nourish the soil.

MUNICIPAL COLLECTION Collection of solid wastes by a public agency.

MUNICIPAL SOLID WASTE Includes non-hazardous waste generated in households, commercial and business establishments, institutions, and light industrial wastes; it excludes industrial process wastes, agricultural wastes, mining wastes, and sewage sludge.

NEWSPRINT A generic term used to describe paper of the type generally used in the publication of newspapers.

NIMBY Acronym for "Not In My Backyard"; expression of opposition to the siting of a facility based on the particular location proposed.

NONFERROUS METAL Metals which contain no iron, such as aluminum, copper, brass, and bronze.

ORGANIC WASTE Waste material from substances composed primarily of chemical compounds of carbon in combination with other elements, primarily hydrogen. These materials include paper, wood, food wastes, plastics, and yard waste.

PACKER TRUCK Type of solid waste collection vehicle used for residential collection that compacts refuse into high density masses for maximum collection efficiency. Can incorporate a rear loading or top loading device.

PAPER The name for all kinds of matted or felted sheets of fiber formed on a fine screen for a water suspension.

PARTICIPATION RATE A measure of the number of people participating in a recycling program compared to the total eligible.

PATHOGEN An organism capable of producing disease.

PET Polyethylene terephthalate, a plastic resin used to make packaging, commonly used to make plastic soft drink bottles.

PLASTICS Non-metallic compounds that result from a chemical reaction, and are molded or formed into rigid or pliable construction materials and fabrics.

POLYPROPYLENE Heavy-duty plastic.

POLYSTYRENE A hard, dimensionally stable thermoplastic that is easily molded.

POLYVINYL CHLORIDE A common plastic material which is tasteless, odorless, and generally insoluble; acronym is PVC.

POST CONSUMER RECYCLING The reuse of materials generated from residential and commercial waste, excluding recycling of material from industrial processes that has not reached the consumer, such as glass broken in the manufacturing process.

PRICE PREFERENCE A means by which an incentive is provided to purchase recycled goods even if they are more expensive than non-recycled goods.

PULPING The operation of reducing a cellulosic raw material into a pulp suitable for further processing into paper.

RECYCLED MATERIAL Material that can be utilized in place of a raw or virgin material in manufacturing a product and consists of materials derived from post-consumer waste, industrial scrap, material derived from agricultural wastes or other items, all of which can be used in the manufacture of new products.

RECYCLING Specifically separating a given waste material from the waste stream and processing it so that it may be used again as a raw material for products which may or may not be similar to the original.

RECYCLING CENTER A place where people bring items to be recycled.

RESIDENTIAL WASTE Waste materials generated in single and multiple family homes; when multiple family units exceed four, these wastes are usually collected in large containers by commercial haulers.

RESIDUE Materials remaining after processing, composting, and recycling.

RESOURCE RECOVERY A general term used to describe the extraction of materials or energy from wastes.

REUSE The return of a commodity or product into the economic stream for use in exactly the same form and kind of application as before, without any change in its identity.

ROLL-OFF CONTAINER A steel box with wheels used to collect waste at a site, such as a construction site, that can be rolled onto a truck using a winch and then taken to a disposal facility for discharge. The empty container can then be trucked to another site and rolled off the truck for stationary waste collection.

RUBBER A natural or synthetic elastic material comprised of polymers. Chemical treatment can enhance properties required for tires, shoes, insulation, and other products.

RUBBISH Non-putrescible solid wastes, including ashes, consisting of both combustible and noncombustible materials such as paper, cardboard, tin cans, wood, glass, bedding, crockery or litter of any kind.

SCALEHOUSE An building located at the entrance of a recycling or disposal facility where weigh scales are placed.

SCAVENGER One who illegally removes materials at any point in the solid waste system. An alternate name for waste hauler or carter.

SCRAP Discarded or rejected industrial waste material suitable for reprocessing.

SECONDARY MATERIALS Materials that are used in place of a primary or raw material in manufacturing a product, often handled by dealers and brokers in "secondary markets".

SHREDDER A size reduction machine which tears or grinds materials to a smaller and more uniform particle size. Shredding processes are also called size reduction, grinding, milling, comminution, pulverization, hogging, granulating, breaking, macerating, chipping, crushing, cutting, rasping.

SMALL QUANTITY GENERATOR A generator who produces less than 100 kilograms of hazardous waste per month (or accumulates less than 100 kilograms at any one time) or one who produces less than 1 kilogram of acutely hazardous waste per month (or accumulates less than 1 kilogram of acutely hazardous waste at any one time).

SOLID WASTE A general term for discarded materials destined for disposal, but not discharged to a sewer or the atmosphere. Solid waste(s) can be composed of a single material or a heterogeneous mix of various materials including semi-solids.

SOLID WASTE DISPOSAL Disposal of all solid wastes through landfilling, incineration, composting, chemical treatment, and any other method which prepares solid wastes for final disposition.

SOLID WASTE MANAGEMENT A planned program for effectively controlling the generation, storage, collection, transportation, processing and reuse, conversion or disposal of solid waste in a safe, sanitary, aesthetically acceptable, environmentally sound and economic manner. It includes all administrative, financial, environmental, legal, and planning functions, as well as the operational aspects of solid waste handling and resource recovery systems.

SOURCE REDUCTION The design, manufacture, acquisition, and reuse of materials including products and packaging, so as to minimize the quantity and/or toxicity of waste produced. Source reduction prevents waste either by redesigning products or by otherwise changing societal patterns of consumption, use, and waste generation.

SOURCE SEPARATION The segregation of specific waste materials at the point of discard for separate collection.

SPECIAL WASTES Hazardous wastes by reason of their pathological, explosive, radioactive, or toxic nature.

SPECIFICATION Clear and accurate description of the technical requirement for materials, products or services. Specifies the minimum requirement for quality and construction of materials and equipment necessary for an acceptable product in the form of written descriptions, drawings, commercial designations, industry standards, and other descriptive references.

STATIC PILE SYSTEM A windrow composting method in which air ducts are generally installed under or in the base of compost piles so air can be blown into or drawn through the pile.

SUBTITLE D That portion of RCRA dealing with non-hazardous solid waste treatment, storage, and disposal facilities.

THERMOPHILES Bacteria or other microorganisms which grow best at temperatures of roughly 45 to 60 degrees C.

TIN CAN Made from tin-plated steel.

TIPPING FEE The charge to unload waste materials at a transfer station, processing plant, landfill, or other disposal site.

TIPPING FLOOR Unloading area for vehicles that are delivering waste materials to transfer stations, incinerators, or other processing plants.

TONS PER DAY Usually referring to the capacity of a paper mill, refuse processing plant, landfill, etc.

TRANSFER TRAILER A vehicle used to transport large quantities of waste over long distances.

TRANSFER STATION Supplemental transportation systems, an adjunct to route collection vehicles to reduce haul costs or add flexibility to the operation. Typically route vehicles empty into a large hopper from which large semi-trailers, railroad gondolas, or barges are filled. There may be some compaction of refuse. Transfer stations may be fixed or mobile, since the larger compacting collection vehicles can serve this function.

TRASH Material considered worthless, unnecessary or offensive that is usually thrown away; generally defined as dry waste material, but in common usage it is a synonym for garbage, rubbish, or refuse.

TROMMEL A large revolving cylindrical screen used as a waste separation technique.

TUB GRINDER Machine used to grind or chip wood wastes for mulching, composting, or size reduction.

VARIABLE CAN RATE A charge for solid waste services based on the volume of waste generated measured by the number of containers set out for collection.

VEGETATIVE WASTES Plant clippings, prunings, and other discarded material from yards and gardens. Also known as yard rubbish or yard waste.

VOLUME REDUCTION The processing of waste materials so as to decrease the amount of space the materials occupy, usually by compacting or shredding (mechanical), incineration (thermal), or composting (biological).

WASTE Useless, unwanted, or discarded material resulting from natural community activities. Wastes include solids, liquids, and gases. Solid wastes are classified as refuse.

WASTE EXCHANGE A computer and catalog network to redirect waste materials back into the manufacturing process by matching companies generating specific wastes with companies that use those wastes as manufacturing inputs.

WASTE PROCESSING An operation such as shredding, compaction, composting, or incineration, in which the physical or chemical properties of wastes are changed.

WASTE PAPER Any paper or paper product which has lost its value for its original purpose and has been discarded. The term is most commonly used to designate paper suitable for recycling, as paper stock.

WASTE REDUCTION The practice of producing smaller quantities of disposable waste. Waste reduction usually entails changing manufacturing processes and packaging practices to foster more recycling and less dependency on disposable goods.

WASTE STREAM The waste output of a region, community, or facility.

WHITE GOODS Discarded kitchen and other large, enameled appliances, as washing machines and refrigerators.

WHITE PAPER Printers term of unprinted paper, even if colored.

WINDROW Composting material stacked in a triangular prism shape.

WINDROW SYSTEM Composting technique in which waste is placed in either aerated static piles or turned, windrowed piles to digest.

WOOD FIBER Elongated, thick-walled cells of wood, commonly called "fiber".

WOOD PULP A fibrous raw material derived from wood for use in most types of paper manufactured by mechanical or chemical means both from hardwood and softwood trees.

YARD WASTE Leaves, grass clippings, prunings, and other natural organic matter discarded from yards and gardens.

REFERENCES

1. Apotheker, Steve, Reverse Vending: A New Market Opportunity, Resource Recycling, October, 1993, pp. 54-61.

2. Apotheker, Steve, Participation in Drop-Off Recycling Programs, Resource Recycling, January 1991, pp. 70-77.

3. Arthur D. Little, Inc., A Report on Advance Deposit Fees, Cambridge, MA, 1991.

4. Artz, Nicholas S., and E. Robert Yoos, Composting Potential in Municipal Solid Waste Management, Prairie Village, Kansas: Franklin Associates, Ltd., 1990.

5. Baker, Stephen, Suddenly, There's Aluminum Everywhere, Business Week, October 25, 1993, p. 46.

6. Bogert, Susan and Jeffrey, The Economics of Recycling, Resource Recycling, September 1993, pp. 76-80.

7. Broughton, Anne, C., MRFS Go High Tech, Recycling Today, March, 1993, pp. 48-57.

8. Broughton, Anne, C., State Recycling Laws: Help or Hindrance?, Recycling Today, February, 1993, pp. 121-130.

9. Bullock, Dave and Debbie Burk, Commingled vs. Curbside Sort, BioCycle, June 1989, pp. 35-36.

10. Cabaniss, Amy, D., Recycling education: the critical link, Resource Recycling, February, 1993, pp. 62-65.

11. California Assembly, Assembly Bill 939, October 1991.

12. Center For Plastics Recycling Research, Market Research on Plastics Recycling, Piscataway, New Jersey: Rutgers, The State University of New Jersey, 1990, Technical Report #31.

13. Chertow, Marian, Garbage Solutions: A Public Official's Guide to Recycling and Alternative Solid Waste Management Technologies, Washington D.C.: National Resource Recovery Association and the United States Conference of Mayors, 1989.

14. Clarke, Marjorie J., The Paradox and the Promise of Source Reduction, Solid Waste and Power, February 1990, pp. 38-44.

15. Cole, Michael A., and Karen K. Leonas, Degradability of Yard Waste Collection Bags, BioCycle, March 1991, pp. 56-63.

16. Commonwealth of Massachusetts, A Financial Effect Determination of Mandatory Recycling on Massachusetts Cities and Towns, Boston: Office of the State Auditor, June 1992.

18. Congress of the United States, Office of Technology Assessment, Facing America's Trash: What Next For Municipal Solid Waste, Washington, D.C., 1989.

18. Couling, Thomas, The Pros and Cons of Composting, American City and County, December 1990, pp. 56-62.

19. Crampton, Norm, Full Cost Accounting: What Is It? Will It Help Or Hurt Recycling, Resource Recycling, September, 1993, pp. 57-61.

20. Deisch, Mitchell, Mt. Pleasant "Goes Green" A User Fee Success Story, December 1989.

21. Environmental Defense Fund, Coming Full Circle - Successful Recycling Today, New York, 1988.

22. Fishbein, Bette K. and Caroline Gelb, Making Less Garbage - A Planning Guide for Communities, New York: INFORM, 1992.

23. Flavell, J.W., A Tribute to Piaget, Society for Research in Child Development Newsletter, The Society for Research in Child Development, Chicago, Illinois, Fall, 1980.

24. Garino, Robert J., UBC Recycling on a Roll, Scrap Processing and Recycling, May/June 1991, pp. 115-117.

25. Gatton, David, Municpal Waste Management Association, Interview, 1993

26. Geller, E.S., and Galen R. Lehman, Motivating Desirable Waste Management Behavior: Applications of Behavior Analysis, Journal of Resource Management and Technology, December 1986, pp. 65-67.

27. Glenn, Jim, An Industry Shapes Up For Recycled Plastics, BioCycle, January 1991, pp. 38-67.

28. Goldman, Matthew, What Do Those Recycling Numbers Mean, The Weston Way, Fall 1990.

29. Grove, Christy, Demystifying the Curbside Plan, BioCycle, June 1989, pp. 45-47.

30. Haverty, Tamar, E., and Richard Regan, Recycling: Equal Access to Capital, Resource Recycling, October 1993, pp. 75-80.

31. HDR Engineering, Inc., Muscatine County Conceptual Solid Waste Management and Recycling System, Prepared For Muscatine, Iowa, September 1993.

32. HDR Engineering, Inc., Public Opinion Survey Findings: Monmouth County, New Jersey, Municipal Solid Waste Management Program, May 1993.

33. HDR Engineering, Inc., Yard Waste Collection and Composting Options, Prepared For the El Dorado Hills Community Service District, October 1992.

34. HDR Engineering, Inc., Union County Recycling Plan Update Study, Union County Utilities Authority, June, 1991, pp. 61-149.

35. HDR Engineering, Inc., Technology Overview and Evaluation, Prepared for the South Plains Association of Governments, December 1991.

36. HDR Engineering, Inc., Pilot Yard Waste Collection Study, Prepared for Orange County, Florida, Orlando, September 1991.

37. HDR Engineering, Inc., High-Rise and Multi-Family Recycling Feasibility Study, Prepared For the Northern Virginia Planning District Commission, July 1990.

38. Hickman, Lanier H., The Solid Waste Association of North America, Interview, December, 1993

39. Hickman, H. Lanier Jr. and Eric G. Reimers, Unit Pricing For Municipal Solid Waste Management Services, International Directory of Solid Waste Management 1993/4 The IWSA Yearbook, November 1993, pp.22-22.

40. Hickman, H. Lanier Jr., Looking At Municipal Solid Waste Composition and Mandatory Removal Rates in North America, Silver Spring, MD: Solid Waste Association of North America, 1991.

41. Hunsicker, Mark, The Rest of the Story, Recycling Today, July 1992, pp.48-54.

42. Kelly, Scott, Large Scale Yard Waste Composting, BioCycle, September 1993, pp. 30-32.

43. Kovacs, William L., Martha E. Pellegrini, Flow Control: The Continuing Conflict Between Free Competition and Monopoly Public Service, Washington, D.C.: Resource Recovery Report, 1993.

44. Kraten, Steven, Market Failure and the Economics of Recycling, Environmental Decisions, April 1990, pp. 20-25.

45. Kropf, Barbara and Caroline Mixon, Drop-Off Recycling in Tampa: Maximizing Efficiency and Quality, Resource Recycling, January 1993, pp. 51-56.

46. Littleton, George, Recycling For Success, Envirosouth, Spring 1990, pp. 15-17.

47. McEntee, Ken, Paper-Makers Pushed to Do More Recycling, Recycling Today, February, 1990, pp. 120-126.

48. McKinstry, R.B., and W.M. Prendergast, Drafting a Mandatory Recycling Ordinance, BioCycle, June 1989, pp. 78-80.

49. Mersky, Ronald L., Designing a Recycling Program, Public Works, May 1989, pp. 67-69.

50. Mishkin, Andrew, Beveridge & Diamond, P.C., Statement quoted in the Minutes of the Recycling Forum, January, 1992.

52. Monmouth County, New Jersey, Monmouth County Public Opinion Survey, May 1993.

52. Murdoch, Joseph D., Robert J. Collins, and John D. Williams, Mandatory Recycling- Five Years Later, MSW Management, November 1992, pp. 44-49.

53. National Recycling Coalition, Inc., State of Recycling - 1993, Washington, D.C., August 1993.

54. National Recycling Coalition, Inc., State of Recycling - 1993, Washington, D.C., October 1993.

55. National Solid Waste Management Association, The Cost to Recycle At a Materials Recovery Facility, Washington, D.C., 1992.

56. O'Brien, Jeremy K. and John F. Williams, Calculating Your Maximum Recycling Potential, Environmental Decisions, December 1990, pp. 19-21.

57. O'Brien, Jeremy K., Integrated Collection: The Key to Economical Curbside Recycling, Solid Waste and Power Magazine, August 1991, pp. 62-70.

58. Oneida-Herkimer Solid Waste Authority, Utica, New York, Local Solid Waste Management Plan, 1990, p.10.

59. Pferdehirt, Wayne, P., P.E., AICP, Planning Bigger, Faster, More Flexible MRFs, Solid Waste & Power, October, 1990, pp. 53-60.

60. Phipps, Harry, Scrap Aluminum Lowers Sheet Raw Material Costs, Resource Recycling, March 1992

61. Plastics Recycling Foundation, Plastics Recycling: An Overview, Washington, D.C., 1990.

62. Plastics Recycling Foundation, Plastics Recycling: From Vision to Reality, Washington, D.C., 1990.

63. Platt, Brenda and Jill Zachary, Co-Collection of Recyclables and Mixed Waste: Problems and Opportunities, Washington, D.C.: Institute For Local Self-Reliance, September 1991.

64. Porter, J. Winston, Dr., Status Report on Municipal Solid Waste Recycling, The Solid Waste Task Force, Government Programs Steering Committee, American Institute of Chemical Engineers, July 1992

65. Powell, Jerry, The Ups and Downs in Bottle to Bottle Plastics Recycling, Resource Recycling, May 1992

66. Powell, Jerry, Recycling and the Law: The Flow Control Battle, Resource Recycling, September, 1993, pp. 35-38.

67. Powell, Jerry, The Decline of the Legislator, The Rise of The Regulator: Recent Trends in State Recycling Programs, Resource Recycling, October, 1992, pp. 44-45.

68. Powell, Jerry, The Common Recyclable: The Growth in Plastic Bottle Recovery, Resource Recycling, January 1991, pp. 40-46.

69. Powers, Roger, Keep America Beautiful Inc., Interview, December, 1993

70. Rapkin, Andrew, Aluminum Cans: Market Dynamics, Resource Recycling, October, 1993, pp. 32-39.

71. Ragsdale, James V., Peter Stasis, Michael J. Rudd, and Joan Bradshaw, Mulch Production From Yard Trash, BioCycle, September 1992, pp. 34-37.

72. Resource Integration Systems, Ltd., Public Sector Commercial Recycling Programs: 10 Case Studies, Prepared for the City of Los Angeles, Office of Integrated Waste Management, June 1990.

73. Rhea, Marsha, National Recycling Coaltion, Inc., Interview, December, 1993

74. Rogers, Fred, Mister Rogers Activities For Young Children About the Environment and Recycling, Pittsburgh, Pennsylvania: Family Communications, Inc., 1990.

75. Rogoff, Marc J., and John F. Williams, The Waste Hurdles, American City and County, January 1993, pp. 46-50.

76. Ruston, John, Developing Recycling Markets for the Components of Residential Mixed Paper, Resource Recycling, January 1992.

77. Santrock, J.W. and S.R. Yussen, Child Development and Introduction, Wm. C. Brown Publishers, Dubuque, Iowa, 1989, pp. 13, 149, 168-169, 298.

78. Scarlett, Lynn, Recycling Costs: Clearing Away Some Smoke, Solid Waste and Power, August 1993, pp. 12-17.

79. Schickendanz, J.A., K. Hansen, and P.D. Forsyth, Understanding Children, Mayfield Publishing Company, Mountain View, California, 1990, pp. 60-68.

80. SCS Engineers, Inc., Final Report: Kent County, Delaware Waste Composition Study, Prepared For the Delaware Solid Waste Authority, August 1990.

81. SCS Engineers, Inc., First Seasonal Report, Waste Composition Study at the Northern Solid Waste Management Center, New Castle, Delaware, Prepared For the Delaware Solid Waste Authority, November 1990.

82. Selby, Mark D., Yard Waste Collection, BioCycle, June 1989, pp. 52-54.

83. Sherman, Stratford P., Trashing a $150 Billion Business, Fortune, August 28, 1989, pp. 92-98.

84. Slovin, Jill, Developing Markets to Close the Loop, World Waste, October 1993, pp. 42-46.

85. Solid Waste Association of North America, Policy Positions on Municipal Solid Waste Management, Silver Spring, MD, December 1993.

86. Solid Waste Association of North America, Manager of Landfill Operations Training Course Manual, Silver Spring, MD, May 1992.

87. Solid Waste Association of North America, Compendium of Solid Waste Management Terms and Definitions, Silver Spring, MD, 1991, Publication #GR-G 0001.

88. Solid Waste Association of North America, Policy T-9, The Role of the Public Sector in the Management of Municipal Solid Waste, Silver Spring, MD.

89. Solid Waste Authority of Palm Beach County, Florida, Palm Beach County Recycling Program, West Palm Beach, Florida, October 1989.

90. Sound Resource Management Group, Inc., The Economics of Recycling and Recycled Materials, Prepared for the Clean Washington Center, Seattle, Washington, June 1993.

91. Trombly, Jeanne, Developing Non-Traditional Glass Markets, Resource Recycling, October 1991.

92. U.S. Conference of Mayors, U.S. Solid Waste Composting Facility Profiles, Washington, D.C., May 1992.

93. U.S. Conference of Mayors, Recycling in America: Profiles of the Nation's Resourceful Cities, Washington, D.C., June 1991.

94. U.S. Environmental Protection Agency, Decision-Makers Guide to Solid Waste Management, Washington, D.C., EPA/530-SW-89-072, November 1990.

95. U.S. Environmental Protection Agency, Variable Rates in Solid Waste: Handbook For Solid Waste Officials, Volume I - Executive Summary, EPA/530-SW-90-084A, September 1990.

96. U.S. Environmental Protection Agency, Charging Households For Waste Collection and Disposals: The Effects of Weight or Volume-Based Pricing on Solid Waste Management, Washington, D.C., EPA/530-SW-90-047, 1990.

97. Waste Management, Inc., Recycling in the 90's, Oak Brook, Illinois, 1992.

98. Watson, Tom, Reverse Vending Revisited: Great Losses and High Hopes, Resource Recycling, March 1990, pp. 26-77.

99. Williams, John, Beginning with Preschoolers: A Recycling Education Initiative, Air & Waste Management Association, Preceedings of the Annual Meeting, June 1991.

100. Williams, John, Mega-MRFs & Full Service Contracts: A Key to Public/Private Ventures, MSW Management, October 1992.

101. Williams, John, Municipal Solid Waste Composting: Looking After the Community's Interests, Air & Waste Management Association, Proceedings of the Annual Meeting, June 1992.

102. Williams, John, and Ann Kulik, American's Attitudes About Recycling, American City and County, July 1992, pp. 48-50.

103. Wood, James J., The Challenge of Multi-Family Recycling, Resource Recycling, June 1991, pp. 34-40.

INDEX